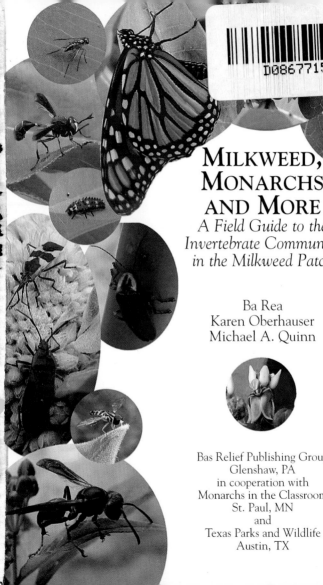

MILKWEED, MONARCHS AND MORE
A Field Guide to the Invertebrate Community in the Milkweed Patch

Ba Rea
Karen Oberhauser
Michael A. Quinn

Bas Relief Publishing Group
Glenshaw, PA
in cooperation with
Monarchs in the Classroom
St. Paul, MN
and
Texas Parks and Wildlife
Austin, TX

This project was supported in part by the
National Science Foundation.

*Opinions expressed are those of the authors
and not necessarily those of the Foundation.*

*All photographs in this publication are the property of the original photographers.
Please see the image credits on pages 3 and 4.*

© Ba Rea 2003

All Rights Reserved.
Published by
Bas Relief Publishing Group
P.O. Box 426, Glenshaw, PA 15116
http://www.basrelief.org

ISBN 0-965-7472-2-0
Library of Congress Control Number: 2003090447

ACKNOWLEDGEMENTS

The authors of **Milkweed, Monarchs and More** would like to express their gratitude to the many people whose expertise, images and enthusiasm contributed to the creation of this field guide. Various scientists assisted us in our efforts to search out and identify arthropods, many people contributed images and others offered us their insights on the format and language of the field guide. We especially like to express our gratitude to Patrick Dailey of Lewis and Clark Community College, who contributed a large number of excellent images and to Cindy Petersen and Jolene Lushine who have carefully edited the text and created the glossary, index and table of contents.

All photographs in this publication are the property of the original photographers. The contributors are listed below. The two-letter code preceding each name is used in the Image Credit listing on page 4 to identify the owner of each image. [i.e. The notation "P. 28 - AA 58; PD 59, 60" translates as: On page 28 image 58 belongs to Anurag Agrawal and images 59 and 60 belong to Patrick Dailey.]

IMAGE CONTRIBUTORS

AA	Anurag Agrawal, Department of Botany, University of Toronto, Toronto, Ontario.	EH	Elizabeth Howard, Journey North.
HA	Harlen E. Aschen, Port Lavaca, Calhoun Co, TX.	MW	Monarch Watch, Kansas University, Lawrence, KS.
CB	Charles Brown, Arvada, CO.	KO	Karen Oberhauser, Monarchs in the Classroom, University of Minnesota, St. Paul, MN.
CC	Carol Cullar, Rio Bravo Nature Center, Eagle Pass, TX.		
JC	J. K. Clark, courtesy of University of California Statewide IPM Program.	TP	Tim Pohl, Winnipeg, Manitoba.
		JP	Jackie Pfeiffer, Butler, PA.
		MQ	Michael A. Quinn, Texas Parks and Wildlife, Austin, TX.
PD	Patrick J. Dailey, Lewis and Clark Community College, Godfrey, IL.	LR	Linda Rayor, Dept. Entomology, Cornell University, Ithaca, NY.
BD	Bastiaan (Bart) Drees, TX.	BR	Ba Rea, Bas Relief Publishing Group, Glenshaw, PA.
IG	Ilse Gebhard, Kalamazoo, MI.		
JG	Jacalyn Loyd Goetz, Overland Park, KS.	WS	William Schiff, Schiff Printing, Pittsburgh, PA.
CJ	Carole Jordan, Laurel, MS.	LS	Lynn Scott, Dunrobin, ON.
JH	Dr. John Haarstad, Cedar Creek Natural History Area, University of Minnesota, East Bethel, MN.	LY	Lisa Young, Louisville, CO.

IMAGE CREDITS

Cover - Front: BR all images; Back inside and out: CJ both images.

P. 1, 6, 7, 8 - BR all images

P. 9 - HA 8; BR 5, 6, 7

P. 10 - MQ 9, 12; BR 10, 11, 13, 14

P. 11, 12 - BR all images

P. 13 - KO 18; BR 19, 20

P. 14, 15, 16, 17 - BR all images

P. 18 - KO 30; HA 32; BR 31, 33, 34

P. 19 - BR 35

P. 20 - LY 36; HA 37; KO 38

P. 21 - JP 39; BR 40, 41

P. 22 - CC 44; BR 42, 43

P. 23 - CC 45a, b; MW 46

P. 24 - PD 47; BR 48

P. 25 - PD 49, 50b, c, f, h; AA 50g; BR 50a, d, e, i

P. 26 - PD 52a; BR 51, 52b, c, 53, 54

P. 27 - AA 55; BR 56, 57

P. 28 - AA 58; PD 59, 60

P. 29 - AA 62; PD 61, 64; HA 69; BR 63, 65-68, 70, 71

P. 30 - PD 72, 73

P. 31 - PD 74; BR 75; AA 76; MQ 77

P. 32 - PD 78, 79, 83; BR 80, 82

P. 33 - MQ 84; JG 85; PD 86

P. 34 - BR 87; PD 88, 89

P. 35 - BR 90, 91, 92, 93

P. 36 - MQ 94, 96; PD 95; BR 97

P. 37 - MQ 98a, b; PD 99; BR 100, 101

P. 38 - KO 102; AA 103

P. 39 - PD 104d-k; WS 104a, b

P. 40 - PD 105, 107, 108; BR 106

P. 41 - PD 108, 110-113; BR 109

P. 42 - PD 114; BR 115a - c; WS 115d, e, f

P. 43 - PD 116; AA 117

P. 44 - MQ 118; JC 119; PD 120

P. 45 - PD 121, 123, 125; MQ 122, 124

P. 46 - AA 126; PD 128, 130, 132 - 134; BR 127, 131, 135-137

P. 47 - PD 138, 141; JH 139; BR 140, 142

P. 48 - PD 143; BR 144, 146; MQ 145; JC 147

P. 49 - PD 148, 149, 153, 156, 158; MQ 150-152, 157; BR 154

P. 50 - JH 159 - 161, 163-165; PD 162, 167

P. 51 - PD 168, 169, 180; BR 170

P. 52 - PD 181, 182; BR 183

P. 53 - BR 184, 186; CB 185; PD 187; MQ 188

P. 54 - PD 189, 190, 192; BR 191, 194; AA 193

P. 55 - AA 195, 197; PD 196

P. 56 - AA 198, 199; PD 200

P. 57 - LY 201; BR 202, 203

P. 58 - BR 204, 205a, b, d, e; LY 205c

P. 59 - BR 206, 209; JH 207, 208, 210-213

P. 60 - JH 214, 215, 217-219; BR 216, 220

P. 61 - AA 221; BR 222-225a, b

P. 62 - BR 226

P. 63 - EH 227; MQ 228, 229

P. 64 - MQ 230 - 232; BR 234; JH 233, 235

P. 65 - LY 236; AA 237; PD 238; LS 239

P. 66 - LS 241a; TP 241b; IG 241c; BR 242-246

P. 67 - BR 247; PD 248-251

P. 68 - PD 252, 254; JH 253

P. 69 - MQ 255; PD 256a, b, d, e, 257a, b; BR 256c; 257c

P. 70 - PD 258 - 260; BR 261, 263; MQ 262

P. 71 - CB 264; PD 265, 266

P. 72 - JC 257; PD 258

P. 73 - BR 269a, c; PD 269b; MQ 270

P. 74 - PD 271; BR 272; JH 273

P. 75 - JH 274-276

P. 76 - BR 277; JH 278, 279b; PD 279a

P. 77 - LR 280; JH 281

P. 78 - JH 282, 283

P. 79 - BD 284; BR 285, 286

P. 80 - BR 287; JH 288

P. 81 - PD 289, 290

P. 82 - JH 291, 292; BR 293; PD 294

P. 83 - BR 295, 297; PD 296, 298

P. 84 - MQ 299, LR 300

P. 85 - BR 301 a, c, 302, 303; LR 301 b

P. 86 - AA 304; BR 305, 306; PD 307

P. 87 - BR 308; KO 309

CONTENTS

In the Field

Careful observation of the milkweed community and the interactions of its diverse members provides a fascinating glimpse into the complex interdependence of living organisms. The approximately 110 species of milkweed found on the North American continent grow under a wide variety of conditions, each with its own complement of invertebrates. Many insects are attracted to the nectar and pollen of milkweed flowers. Others feed directly on milkweed leaves, seeds, stems or roots. Predatory insects and arachnids are attracted to the plentiful prey population. Many species may be found temporarily resting on the plants.

If you are new to the milkweed patch, these suggestions may help you get started with your field observations. Wear long pants to protect your legs from briars, grass cuts and any biting insects that might be present. Bring along a hand lens, a notebook and pencil, and small containers (film canisters work well) for collecting interesting insects and other finds to further investigate at home. As you get more involved in your observations, a camera will help you record what you see and close focusing binoculars will allow you to get a close-up view of insects' behavior without getting close enough to disturb them.

Walk quietly and stop frequently. Stop for several minutes at a time to watch the activity on one milkweed plant. Species scared off by your approach may return if you are still. Scan the leaves, stems and flowers for unusual shapes, color or activity. Take your time. It may take a few minutes of concentration to spot milkweed community activity. Watch for signs like chewed leaves and frass (caterpillar droppings). Be alert for movement. Keep notes on what you see and when you see it. Once you have found an interesting milkweed community member, watch for it again. Observe its interactions and changes. Visit the milkweed patch at different times of day. Note the effects that time of season, temperature and weather conditions, as well as the age and condition of the plants, have on the activity you see.

1. Milkweed beetle

Using This Field Guide

Due to the constraints of space and knowledge, we have limited this field guide to the orders and primary families of insects and other arthropods found on milkweed in North America north of the Mexican border. In this text the term "North America" is used to cover only that area.

Identification of insects often requires careful observation of wing cells, placement of hairs, **apical** spurs and other details. The information provided here is general and intended to present an overview of the variety of behavior and structure of the arthropods in a milkweed community. More detailed guides may be required to make definitive identifications.

Green bands on the outside edges of the pages of the field guide summarize in a few words what each page covers.

2. Jumping spider with fly

There are four main sections in this guide:

- **Investigating and Sharing What You Find** covers terms and concepts useful for organizing and sharing your observations.
- **Milkweed** is an overview of Asclepiadaceae, the milkweed family, including a photographic sampling of a few species found in North America.

Role Codes

HERB	Herbivore
MW	Herbivore that eats milkweed
NECT	Nectivore
PRED	Predator
PARA	Parasite
DEC/S	Decomposer or Scavenger
PASS	Passerby

- **The Arthropods** has two sections. The first, **Arthropod Body Basics,** is an overview of insect and arachnid body structures. The second, **Arthropods of the Milkweed Community,** provides entries on arthropods at the order and family levels. Entries include a photo of one or more representative individuals, with scientific and common names, identifying features, life cycles, size, and range. Color coded symbols, located on the outside edge of the pages, are included for each entry in this section. They alert you to the animals' roles in the milkweed community. (*See key on left.*)

- **Glossary, References and Index**. A glossary of terms (including all bolded words in the body of the text), a short bibliography of a few of the most important references used in the creation of this text, and an index for searching this volume are included at the end of the book.

7

Investigating and Sharing What You Find

The milkweed community is a stage for a season-long series of dramas involving a large cast of fascinating characters. Creatures hatch and grow. Some nestle into the foliage intent on eating. Others glide down onto blossoms to gather nectar. Predators lurk and ambush. An apparent nectivore may be watching for an opportunity to lay an egg, from which its parasitic larva will hatch, on an appropriate host. This section introduces words and concepts that can help you organize, think and talk about what you see in the milkweed patch.

The Food Web and the Roles of the Members of the Milkweed Community

An exchange of energy and materials—collected, stored and passed between organisms—is the basic interaction underlying much of the activity in any natural community. This series of interactions is often referred to as a **food chain**, but the many connections each organism has with others in the community make the term **food web** more accurate. Recognizing and understanding organisms' roles in the food web will help you to interpret the story playing out in your milkweed patch.

Symbols can be found in this section on the photos used as examples for each role. Although the creatures in the examples often have multiple roles in the community, only the symbol for the role being illustrated will be present in this section. Throughout the field guide these symbols indicate the roles for each entry.

The energy that circulates through a living community on our planet comes from the sun. It enters the community through green plants.

Green plants are referred to as **producers**. **Chloroplasts —organelles** inside the cells of green plants and some bacteria—capture sunlight and store that energy in bonds in sugar molecules. Sugar molecules are the basic food for all living things. Plants also make minerals from the soil available to other living things by incorporating them into their tissues.

4. A. syriaca—Prod

3. Monarch caterpillar

HERB

Animals that eat plants are called **herbivores.** Plants provide the energy they need to live and the materials they need to create their own tissues.

5. Honey bees are important pollinators.

6. Ants drink nectar and do not contribute to pollination.

Nectivores' food also comes directly from plants. Flowers produce nectar to attract them. As they search blossoms for their food, nectivores carry pollen from flower to flower—fertilizing the **ova** to produce seeds for the next generation of plants. However, there are insects that drink the nectar but do not contribute to pollination.

Predators are animals that eat other animals. They use the energy and raw materials stored in the bodies of their prey to provide energy and build their own tissues. Many different adaptations and skills help them capture and consume their prey.

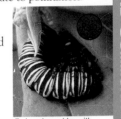

Parasites are organisms that live inside or on other organisms. They may feed on the host's nutrient resources or directly on their living tissues. Some parasites kill their hosts, while others cause varying degrees of harm without actually resulting in the death of their hosts.

7. Jumping spider with monarch caterpillar

A protozoan parasite, *Ophryocystis elektroscirrha* (*O.e.*), has adapted to live and multiply in monarch butterflies' bodies. Although there is evidence that even a mild *O.e.* infection may have health costs for monarchs, the life cycle of this parasite depends on the survival of its host. *O.e.* protozoans develop in monarch larvae and pupae, eventually migrating to the

8. *O.e* spores nestled among scales *O.e.* free scales

tissue that will become the exoskeleton of the adult butterfly. There, they become spores that can withstand exposure to the outside environment. When a female butterfly infected with *O.e.* lays eggs, spores nestled in the scales of her abdomen are left behind for the newly hatched caterpillar to eat. Severe *O.e.* protozoan infestations can be fatal, but in the wild, especially in the north, a caterpillar will likely only encounter the very few spores left behind by its mother. In warmer climates, where the plants do not die back and spores can accumulate on leaves, or in captivity, where many caterpillars are raised together, the number of spores ingested by an early instar caterpillar can reach harmful or fatal numbers.

Parasitoids are insects that live in or on other living animals during their larval stage and usually kill their hosts. Although it is sometimes argued that this is a predatory relationship because the adult or nymph hunts for the host, this field guide classifies them as parasites. Tachinid flies are parasitoids that lay their eggs on many species of caterpillars, including monarchs. The tachinid larvae burrow into the caterpillars' bodies and consume them from within. They eat non-essential tissues first so that the caterpillars live long enough for them to complete their development. Telltale threads hanging from a limp caterpillar body or chrysalis are a sign that tachinid larvae have left their host to pupate on the ground. (*See page 70 for more on tachinid flies.*)

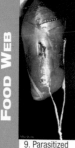

9. Parasitized monarch chrysalis

10. Tachinid fly

Scavengers and **decomposers** are important members of any ecosystem. They consume and break down discarded and dead tissues, making the nutrients in them available to the community again. Many of these organisms can be found under the milkweed plant in the leaf litter and dead grasses. They include insects like termites, which are able, through a complex relationship with bacteria that live in their hind guts, to break down cellulose. Snails and slugs, the crustacean sowbugs and pillbugs, and many types of fungi and bacteria are important scavengers and decomposers as well.

11. Sowbug

13. Earwig

14. Land snail

12. Mayfly

Some insects may be regular visitors to milkweed, especially if the plant is in their habitat, even though they do not consume any part of the plant or interact specifically with any of the other creatures of the community. They can fall victim to some of the milkweed community's predators. We will include some of these creatures as **passersby**.

Biomass is a measure of the volume of living matter in a community. In a sustainable community, producers usually make up the greatest amount of biomass, because at each exchange—from plant to herbivore or herbivore to predator—energy is released and lost to the community through respiration and other processes of living. Most communities can support more herbivores and nectivores than predators and parasites.

Habitats and Communities

The milkweed patch is a **habitat** for many creatures. A habitat is the place (or collection of places) where an organism finds all the things that it needs to survive—food, space (in an appropriate configuration), water, shelter, and air. Insects, which metamorphose, may have very different habitat requirements at different points in their lives.

A **natural community** is the intersection of many habitats. Communities include all of the organisms that live and interact in a particular area. Each community has characteristics created by the organisms involved and many **abiotic** (non-living) features, including weather, microclimates, soil, air quality, water quality and abundance, land formations that affect light or heat, and structures, chemicals or machinery introduced by humans.

In any milkweed community, there are likely to be a number of other plant species. Their interactions and the interactions of the invertebrates that they attract are very much a part of the community. This field guide covers only the milkweed plants. The variety of other plants that can be found in association with various species of milkweed in ecosystems across the continent is beyond our scope, but as you observe your own community, be aware of the impact of the adjacent plants. Nearby plants will affect the mix of invertebrates visiting milkweed plants and may yield clues about the history and future of the milkweed community.

Some milkweed communities will be established in areas that are disturbed regularly. If a disturbed area is being allowed to **naturalize**, you will notice a change in its plant population over time. Common milkweed (*Asclepias syriaca*), the most familiar milkweed species in central and eastern North America, is an early succession plant. It thrives in areas that have been recently disturbed. If the area where it is growing is allowed to go through **natural succession**, common milkweed may eventually be crowded out by plants better equipped to survive in later succession communities. However, there are climax prairie species of milkweed with exactly the opposite problem. *Asclepias meadii* is an endangered **climax** prairie plant. In its natural habitat, it may live to be one hundred or more years old. It is not the encroachment of other plants or herbivores, but the disappearance of the prairie, that threatens its existence.

15. *A. meadii*

11

Each organism in a community has behaviors and structures—adapted over many generations—that optimize its chances for survival. These adaptations affect the organism's developmental **phenology**, movement patterns in time and space, and food consumption. This suite of adaptations, and the resulting use of habitat, is sometimes referred to as an organism's **niche**.

In northeastern and north central North America, both milkweed bugs and milkweed beetles are abundant on common milkweed; however, they occupy different niches in time as well as in their use of the milkweed plant. Milkweed beetle larvae overwinter on milkweed roots. They pupate underground and emerge in early summer as adults. They can be found mating or nibbling on the milkweed leaves and flower buds in June. In late summer, after they have laid their eggs close to the ground on milkweed stems, the number of milkweed beetles visible above ground in the milkweed community decreases dramatically. (*See page 53 for more on milkweed beetles.*)

16. Milkweed beetles

In the same region, milkweed bugs are abundant in late summer. They overwinter as adults and are present throughout the summer. Occasional clusters of adults and nymphs can be seen in June and July, but their numbers grow dramatically in August when milkweed seeds—their preferred food—are ripening. One species of milkweed bugs, *L. kalmii*, has adapted to thrive earlier in the season. In a California study, *L. kalmii* have been documented living as scavengers, predators and herbivores eating non-milkweed plant material, especially in the early summer.[1] The late season *L. kalmii* still congregate on milkweed pods. (*See pages 35-36 for more on milkweed bugs.*)

17. Milkweed bugs

Both monarchs and milkweed tussock moths feed, as caterpillars, on milkweed during the summer in the northeastern and north central United States. These two creatures are **competitors**. They need the same resource—milkweed foliage—but have different **strategies** for optimizing their chances for survival.

---

[1] Root, R. B. 1986. The life of a Californian population of the facultative milkweed bug, *Lygaeus kalmii*, (Heteroptera: Lygaeidae). **Proc. Ent. Soc. Wash.**, 88(2), pp. 201-214.

18. Milkweed tussock moth caterpillars

The milkweed tussock moth lays over fifty eggs in a tight cluster on a milkweed leaf. The caterpillars are **communal feeders**—they hatch and feed together. Their great numbers increase the chances that some from each egg mass will survive to reproduce. This strategy also makes them devastating competitors. As they feed, they eat everything in their path—turning the milkweed plant into a skeleton and devouring any eggs or small caterpillars on the leaves as they go. As they get older they develop thick orange and black, furry tufts. The color and hairs also represent defense strategies. The color announces the toxins that they carry in their bodies from the milkweed they eat and the hairs make them less appetizing.

Monarchs have a very different survival strategy. They lay their eggs one at a time and often only one to a milkweed plant. This way, the caterpillars avoid attention, disease and competition with each other. The 400 or more eggs laid by any one female are spread over a large area ensuring that some will find suitable conditions for survival.

The strategies of living organisms have evolved over many generations and represent features and lifestyles that have given the individuals in each species the best chance for survival.

Classifying and Naming Organisms

In order to communicate about the living things around us, we have named them. However, organisms often have two or more common names and quite frequently very different organisms are given the same common name.

For most kids in America the name "daddy-long-legs" refers to a long-legged, non-spider arachnid common in woodlands and meadows (*left*). But in Europe this same organism is commonly called a harvestman. To further complicate matters, there is a true spider, called a daddy-long-legs spider (*right*), that lives in the dark corners of caves and cellars. Many people refer to the long-legged crane fly as a "daddy-long-legs" as well. Similar confusion surrounds a common name for cicadas. When the early settlers were surprised by a massive

19. Daddy-long-legs, harvestman

20. Daddy-long-legs spider

13

emergence of periodic cicadas, they had no field guides to refer to for information about what was happening, only the Bible. Citing the plague of locusts that Moses' god inflicted on Egypt, they called the cicadas "locusts." Although the biblical locust is a migratory grasshopper, unrelated to the cicadas, the name stuck—and even made it into Webster's Dictionary!

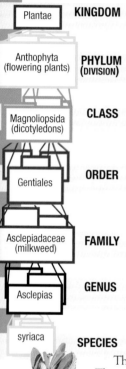

	KINGDOM	
Plantae		Animalia
Anthophyta (flowering plants)	**PHYLUM (DIVISION)**	Arthropoda (jointed foot)
Magnoliopsida (dicotyledons)	**CLASS**	Insecta
Gentiales	**ORDER**	Lepidoptera (scale wing)
Asclepiadaceae (milkweed)	**FAMILY**	Nymphalidae (brush-footed
Asclepias	**GENUS**	Danaus
syriaca	**SPECIES**	plexippus

To make it possible to accurately communicate their observations and research, biologists use a system for classifying living things that was developed in 1759 by Carlos Linnaeus. Called **systematics**, the system is based on the organisms' genetic and evolutionary similarities. All members of a particular grouping are believed to have evolved from common ancestors. There are seven major levels in the system. First, all living organisms are divided into five **kingdoms**: Plants (Plantae), Animals (Animalia), Fungi, Protozoans (Protista) and Bacteria (Monerans). Kingdoms are further divided into **phyla** (plural for phylum). Botanists call groups at this level in the plant kingdom **divisions**. Phyla and divisions are divided into **classes**. Classes are divided into **orders**, orders into **families**, families into **genera** (plural for genus), and genera into **species**. A species is a population of organisms that can successfully interbreed and are in some way different from all other living things.

The naming of living things is called **taxonomy**. The **scientific name** for any organism is made up of its genus name followed by its species name. Scientific names are created to be consistent worldwide, although sometimes names are changed and groupings rearranged as more is learned about organisms.

21. *Asclepias syriaca* 22. *Danaus plexippus*

14

Milkweed

The scientific name for members of the milkweed family is Asclepiadaceae, after Asklepios, the Greek god of healing. The plants of this family have been used for fiber, food and medicinal purposes for ages. Most of the 2000-3000 species in the milkweed family are tropical. The approximately 110 species of milkweed found on the North American continent have quite diverse growth habits, ranging from sturdy, broad-leafed early succession plants to small climax prairie plants and vines. Two features almost all milkweeds have in common are a thick white sap and a unique flower adaptation.

The name "milkweed" comes from the plants' milky sap that contains latex and a toxic alkaloid, known as a cardiac glycoside, which adversely affects birds and mammals. Grazing mammals avoid eating milkweed except when there are no other available green plants. Many insects also avoid them, but some insect species have adapted to live all or part of their lives feeding directly on milkweed. The toxic alkaloids provide protection for these insects by making them toxic to potential predators. The level of toxicity in milkweed species varies widely. Some milkweed species are too toxic even for monarchs.

23. *A. syriaca* umbel

Milkweed flowers vary from bright orange to pink, green or white. The relative sizes of their flower parts and number of blossoms per **umbel** (flower cluster in which the flower stalks arise from a common point) are also quite variable. But unlike most flowers, which distribute their pollen as a powder that shakes off onto nectaring insects, milkweed flowers transfer their pollen in a pair of waxy packets which attach to the legs, proboscises and even the bristly hairs of insects that stop to drink nectar.

On each open milkweed flower, five sepals and five petals fall back against the **flower stalk,** exposing an intricate structure arranged in a five-pointed star pattern. Modified **anthers** form **hoods**, each enclosing a horn and containing large amounts of

24. *A. syriaca* blossom

15

nectar to attract insects. Between adjacent nectar-filled hoods, on the central column of the flower, is a slit, formed by two flaps of modified anther tissue. Inside this slit are found both the **pollinarium**, the structure evolved to package the flower's pollen, and the opening to the **stigmatic chamber** where pollen must be delivered to reach the ovaries and fertilize the milkweed seeds.

A pollinarium is a wishbone-shaped structure. At the top is a clip called a **corpusculum**. The corpusculum can easily be seen as a dark spot on the **flower column**. Two threads, called **rotator arms**, attach two waxy pollen-filled structures, called *pollinia* (singular—pollinium), to the corpusculum. Each pollinium holds enough pollen grains to pollinate an entire milkweed pod.

As a bee, wasp, butterfly or beetle drinks nectar from the flower, its leg (or possibly, some other appendage or body part) gets wedged in the slit. As the insect pulls its leg up, it catches on the corpusculum. If the insect is able to pull its leg free, the entire pollinarium comes out attached to it. Freed from the flower column, the rotator arms on the pollinarium begin to dry. They contract, rotating the pollinia 90 degrees—to the proper position to fit into the slit on the flower column of another milkweed flower. The drying process takes several minutes and represents an interesting adaptation. Most milkweed species are "**self-incompatible**"—they require pollen from a different plant to produce fertile seeds. In the time it takes for the pollinia to rotate into position so that they can fit into the opening to the stigmatic chamber, the insect

Horns

Hoods

Corpusculum

Flower column

Opening to stigmatic chamber

Petals (sepals are hidden underneath)

Flower stalk

25. *A. curassavica*

Pollinarium

Corpusculum

Rotator arms

Pollinia

26. As they dry, the rotator arms turn the pollinia 90°.

can finish nectaring and
move on to the next plant.

Pollinarium

27. Vespid wasp
nectaring on *A.
curassavica* and
collecting pollinia

Once deposited in
the stigmatic chamber,
pollen tubes from individual pollen grains in
the pollinia grow toward the ovules at the
base of the flower. As the pollinator pulls free
once more, it may dislodge another pollinarium
to transport to yet another flower.

28. Bee stuck on
milkweed blossom

Many smaller insects, such as ants,
come to feed without ever moving milkweed pollen. For smaller insects whose
feet do slip into the slits along the flower
column, pollination is a tricky business that
can prove fatal. It is not unusual to find small
bees or butterflies hanging from a milkweed
flower by one leg. Too small to pull themselves
free, they struggle until they die of exhaustion or fall
prey to crab spiders or ambush bugs.

Each successfully pollinated milkweed flower can produce two
pods, but they usually only produce one. When fully developed
and dry, the pod splits open at a single suture to
release seeds attached to white tufts of silk.

A. curassavica

29.
Bud to Pod

NOTE: The proper
botanical term for the
milkweed pod is **"follicle."**

A. syriaca

Other Plants in the Milkweed Community

A diverse plant community is an exciting and beautiful entity. Plants within a community often divide above- and below-ground resources, allowing dozens of species to grow together. Roots extract water and nutrients from different soil depths. Above-ground shoots often grow at different rates, maximizing space and sunlight exposure at different times of the season. It is common for many flowering plants, called **forbs**, to be part of a milkweed community. The types and diversity of these plants will affect the variety of insects attracted to the community.

Several grass species can be found in many milkweed communities as well. Although grasses do produce flowers, it is the wind—not insects—that moves grass pollen from one flower to another. Grasses are an important food source for many herbivores. Their abundance and diversity influence the make-up of the rest of the milkweed community.

Dogbane is a plant that can some- times be found growing in milkweed communities. It is easily mistaken for milkweed. Dogbane plants exude a milky sap when cut, but cannot be eaten by milkweed specific herbi- vores. Dogbane's branching structure and flowers are clues to its identity.

31. Pennsylvania side yard

30. Minnesota soybean field

34. Dogbane

33. West Virginia woodland opening

32. Texas field with bluebonnets

GALLERY OF
MILKWEED

These are just a few of the milkweeds you may encounter in North America. The maps are composites from several sources and are for approximate reference only.

Common Milkweed
Asclepias syriaca

35. *A. syriaca*

Common milkweed is the most prevalent, weedy species of the milkweed family in northeastern North America. It is an early succession, perennial plant growing from a deep **rhizome**.

Stems can be hairy, are usually solitary, and occasionally branched if the main stem has been cut or injured. Multiple stems originating from the same underground rhizome are called **ramets**. They are essentially stalks of the same plant. A plant with many ramets is called a **clone** and often forms dense stands. Mature ramets can be well over one meter (3 feet) tall.

The leaves are opposite, broad and densely haired on the underside. Flowers are purple-tinged and grow in dense to moderately dense, round **umbels** of 20 to 130 blooms. They produce large amounts of nectar and attract many species of pollinators. The pods are covered with soft fleshy "hairs."

Common milkweed is the host plant for 90% of the monarch butterflies that overwinter in Mexico.[2] This species has relatively low toxicity.

A. syriaca

[2] Malcolm, S.B., B.J. Cockrell and L.P. Brower. 1993. Spring recolonization of eastern North America by monarch butterfly: successive brood or single sweep migration? Pages 253-267, in S.B. Malcolm & M.P. Zalucki (eds.), *Biology and Conservation of the Monarch Butterfly.* Natural History Museum of Los Angeles County, Science series 38, Los Angeles.

36. *A. speciosa*

Showy Milkweed
Asclepias speciosa

Showy milkweed grows in clumps along roadways and fields and in open areas. It is the dominant weedy milkweed species of the plains and northwest. The plants are hairy perennials, with stout simple stems and opposite leaves. The hoods on the flowers point upward, making them look like crowns. They grow 1.3 to 1.7 m (4 to 5') tall on prairies in sandy, loamy and disturbed areas from California to British Columbia and central Canada to Texas.

Butterfly Weed
Asclepias tuberosa

37.

38. *A. tuberosa*

Butterfly weed is a perennial with deep, woody rootstalks. The name "tuberosa" refers to knobs in the root system. The leaves are alternate, lance-shaped, smooth on top and velvety under-neath. This milkweed grows in mounds in dry, open soils and has orange to red and sometimes yellow flowers. It has a variety of common names, including pleurisy root, tuberroot and chigger flower and is often cited for its medicinal uses. Butterfly weed is a common garden plant and grows up to 80-90 cm (30-35") tall.

Swamp Milkweed
Asclepias incarnata

Swamp milkweed often grows in clumps with many stalks from short superficial rootstalks. It is found in swamps, wet thickets and on shores in the northeast. It can also survive in much drier habitats and is often cultivated in gardens. Leaves are opposite and stems may be simple or branched. Flowers are small and pink or occasionally white. The pods are narrow and stand upright. Plants can be almost 2 m (6.5') tall.

39. *A. incarnata* pod

40. *A. incarnata* bloom

A. incarnata

Spider Milkweed
Asclepias viridis

Spider milkweed has alternate leaves, often with wavy leaf margins. It is a common Texas milkweed growing on prairies, glades, dry hillsides and pine barrens and is an extremely important food source for the first spring generation of monarchs. It ranges east from Tennessee to Florida and west to Nebraska. Flowers are large, white to green, and showy, usually with only one flower cluster per plant. The plants can be erect or **recumbent**, with stems up to about 60 cm (24") long.

A. viridis

41. *A. viridis*

21

Tropical Milkweed
Asclepias curassavica

Tropical milkweed or blood-flower is not a native of the United States, but has been propagated by humans and is now established as a perennial throughout the south. It is grown as an annual in

42. *A. curassavica*

many northern gardens. It grows from a woody rootstalk, has opposite leaves and small, bright-red to orange flowers with yellow centers. In the south, where it doesn't die back in the winter, tropical milkweed may look like a large bush.

Bluntleaf Milkweed
Asclepias amplexicaulis

A. amplexicaulis is also known as clasping milkweed because of the way that the leaves wrap around the stem. This milkweed prefers sandy soils. The leaves and stems are often purple-tinged. Plants grow up to 80 cm (31") tall.

43. *A. amplexicaulis*

Antelope Horns
Asclepias asperula

A. asperula is a low, stout, herbaceous perennial with many stems clustered around thick rootstalks. The flowers are green. This milkweed is found in sandy and dry conditions. It is an important food source for monarchs in Texas. The plants are usually recumbent, with several stems up to about 60 cm (24") long.

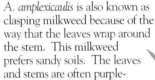

44. *A. asperula*. Note the milkweed bugs in the opening pods in this photo taken in southern Texas in June.

22

45a. *A. oenotheroides* with monarch egg near center

Zizotes Milkweed
Asclepias oenotheroides

45b. *A. oenotheroides* bloom with milkweed bugs

Zizotes milkweed is a low perennial, often only 10-15 cm (4-6") tall, with multiple stems growing from a rootstalk. The leaves are opposite and quite thick with wavy edges. Zizotes milkweed hosts many monarchs in the spring.

Sand Vine
or Honey Vine
Cynanchum laeve

This perennial vine with heart-shaped leaves is one of 36 species of vining milkweeds from the genus *Cynanchum*. Plants from this genus are commonly known as swallowworts. *C. laeve* is an important host for monarchs in the central United States.

Another species of this genus, *C. nigrum*, became established in the 1990s after its intro-

duction to the northeastern United States. While female monarchs will lay eggs on this species, the larvae cannot develop on it.

46. *C. laeve*

23

The ARTHROPODS

The majority of the invertebrates in the milkweed community are from the phylum Arthropoda and from the classes Insecta and Arachnida.

Arthropod Body Basics
INSECTS

Insects have a somewhat hardened outer body wall called an **exoskeleton** that protects their internal organs and prevents desiccation. Insect muscles attached to the inside surface of the exoskeleton make movement possible, in much the same way that muscles attached to our internal skeletons make our movement possible.

Insect bodies are segmented and divided into 3 regions: the **head**, the **thorax** and the **abdomen**. They "breathe" through a system of tubes that bring oxygen directly to their tissues. The tubes open externally at the **spiracles** —usually 2 pairs on the thorax and many on the abdomen. Insects have an open circulatory system with no veins or arteries. A short dorsal tube located in the abdomen functions as a heart, pumping blood, called **hemolymph**, through their bodies.

Insects have many sensory organs located in their body walls. They respond to touch, chemicals, sound, light and vibration. Taste and smell receptors are generally located on the mouthparts, antennae and feet. Sounds are detected through drum-like structures or special hairs sensitive to sound waves. Many insects hear sounds pitched much higher than the sounds we hear.

Head

Thorax

Abdomen

47. Milkweed beetle

Compound eyes

Ocelli

Head

Eyes, antennae and mouthparts are located on the head. Adult insects have two types of eyes: two **compound eyes** on either side of the head and, usually, three simple eyes, called **ocelli**, located on the upper front of the head. The larvae of insects that undergo complete metamorphosis do not have compound eyes.

24

48. Chinese praying mantis

50. Insect heads and legs

Antennae are usually located on the front of the head below the ocelli. Antennae have many different forms and are often used to identify insect groups.

a. Moth

b. Japanese beetle

c. Ladybug beetle

Proboscis

d. Monarch

e. Milkweed bug

Mouthparts are usually located on the anterior or ventral side of the head. There are two general types of insect mouths: chewing and sucking. Chewing mouths have **mandibles** that move from side to side. Sucking mouthparts are modified to create a beak called a **proboscis.** A sucking beak enclosing stylets for piercing is called a **rostrum.** Bees and some other insects have mandibles that move from side to side and a beak-like tongue through which they can suck liquids.

f. Leatherwing beetle

g. Spined soldier bug

Rostrum

Thorax

The **thorax** is the middle region of the insect body. It is divided into three segments: the **prothorax** in the front, the **mesothorax** in the middle, and the **metathorax** at the back. All three segments carry one pair of legs. On most winged insects, the mesothorax and metathorax each have one pair of wings as well. Metathorax wings on flies are replaced by a pair of small modified wings called **halteres.**

Prothorax
Mesothorax
Metathorax

49. Milkweed beetle

h. Hornet

Tongue

Mandibles

i. Insect leg

Coxa
Trochanter
Femor
Tibia
Tarsus

Insect legs are divided into 5 major segments. The segment that attaches to the thorax is the **coxa**. The small segment beyond the coxa is the **trochanter**. The next segment is the **femur**, followed by the **tibia**. The leg ends with a segment called the **tarsus**. The tarsi are the insects' feet. They can have a number of segments and typically end in a pair of claws. The size and shape of the leg segments vary widely and are often used for identification purposes, particularly in beetles.

Insect wings are quite variable and are often used for identification. Most insect order names end in the suffix "ptera," from the Greek word meaning wing and most adult insects have wings. Wing muscles are attached to the walls of the thorax. Their movements are produced by changes in the shape of the thorax.

25

Abdomen

Insect abdomens can have up to ten complete segments, with an eleventh segment generally made up of appendages. Segments are often fused or telescoped together. There are usually no appendages on the abdomen except at the end. Some typical abdominal appendages are feeler-like or clasper-like **cerci** and **ovipositors** (egg laying organs).

51. Above: monarch abdomen. *52. Below:* abdominal appendages; a. earwig, clasper-like cerci, b. praying mantis, feeler-like cerci, c. ichneumon wasp, ovipositor

Metamorphosis

While some females bear live young, most insects begin their lives as eggs. The eggs may be laid singly, in characteristic groupings or within a special casing. A nymph or larva hatches from each egg. As it grows, it gets too big for its exoskeleton and **molts**, shedding its outgrown exoskeleton and replacing it with a newly formed one. Stages between molts are called **instars**. With each molt there are changes in the insect's form as well as in its size. The process is called **metamorphosis**. All insects go through metamorphosis. There are two main types of metamorphosis: simple (or incomplete) and complete.

Simple metamorphosis (*example: milkweed bug, below left*) involves three stages: egg, **nymph** and adult. A nymph usually molts between 4 and 8 times, depending on the species, before becoming an adult. For most insects that go through simple metamorphosis, the immature stage resembles the adult without wings. There are a few exceptions. Mayfly, dragonfly, damselfly and stonefly nymphs are aquatic and breathe with gills. Cicada nymphs live underground, feeding off roots. All of these insects take on adult forms that are very different from their nymph forms.

Complete metamorphosis (*example: monarch below right*) has four stages: egg, **larva**, **pupa** and **adult**. Insects that go through complete metamorphosis have very different forms in their larval and adult stages. After the last larval molt, these insects become pupae. Pupae are usually inactive and do not feed. Some are enclosed in a protective covering, like a cocoon. Adult insects emerge from the pupae.

53. Milkweed bug

54. Monarch

26

ARACHNIDS

Arachnids are wingless arthropods with four pairs of legs. They molt as they grow.

SPIDERS

Spiders have two major body regions. The head and thorax are fused into a region known as the **cephalothorax,** which is joined to the abdomen by a stalk called a **pedicel.** In addition to four pairs of legs, there are two pairs of appendages at the front of the cephalothorax. The first pair, the **chelicerae,** are used for feeding. They can be pincer- or fang-like. The spiders' venom glands open on the chelicerae. The second pair, the **pedipalps,** are used for capturing prey and fertilizing females. Female pedipalps are generally slender and male pedipalps are clubbed.

55. Crab spider
56. Spider body
Pedipalp
Chelicerae
Pedipalp
CEPHALOTHORAX
Pedicel
ABDOMEN

Most spiders have eight simple eyes. The arrangement of their eyes is often important in identifying individual families.

Spider abdomens are not segmented. Silk producing organs, called **spinnerets,** are located on the underside at the rear. All spiders can spin silk, but only some make webs. Young spiders use their silk to disperse through a process called "**ballooning.**" They release long silken threads that they use, like parachutes, to float away.

OPILIONES

The arachnids commonly called daddy-long-legs or harvestmen are not spiders. Their cephalothorax and abdomen are broadly joined forming a compact, oval body. Their abdomens are segmented and their legs are long and slender. The second pair of legs carries sense organs for sensing vibration, tasting and smelling. Opiliones produce neither silk nor venom.

57. Opilione

27

Arthropods of the Milkweed Community

Many kinds of organisms live in the milkweed community, from hummingbirds and frogs to slugs and snails. Sowbugs and pillbugs from the class Crustacea forage in the detritus at the base of the plants and munch on the new growth. However, most of the milkweed community members are arthropods from the classes Insecta or Arachnida. This section introduces insect and arachnid orders with representative families and species likely to be found in the milkweed community.

58.
Fro

INSECTS

PASS

59. Mayfly

ORDER EPHEMEROPTERA
Mayflies

These most primitive winged insects are passers-by in the milkweed community. They may become prey for resident predators if the milkweed stands near a stream or other body of water.

Identifiers: Soft-bodied; wings held above body at rest; 2 to 3 hair-like tails off the end of the abdomen.

Life Cycle: Nymphs are aquatic. Adults live only a day or less and do not eat.

Range: 611 North American species.   **Size:** 3-15 mm (1/8-5/8").

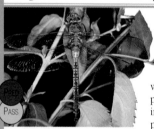

PASS

60. Green Darner, one of the largest, fastest dragonflies in North America

ORDER ODONATA
Dragonflies and Damselflies

Dragonflies and damselflies can be found on milkweed plants, especially near bodies of water. The adults are predators, catching their prey in mid-flight. They may be harvesting flying nectivores around milkweed blooms or simply resting. Larger species will capture and eat butterflies, including monarchs. Adults may travel some distance from water.

Identifiers: Most damselflies rest with wings folded together over their backs; dragonflies rest with wings out to the side.

Life Cycle: Simple metamorphosis. Nymphs are aquatic.

Range: 452 North American species. **Size:** 18-127 mm (3/4-5").

61. Long-horned grasshopper nymph

ORDER ORTHOPTERA
Crickets and Grasshoppers

PASS
HERB

These insects are the familiar crickets, katydids and grasshoppers. They have long hind legs for jumping. They are herbivores but do not usually eat milkweed.

Life Cycle: Simple metamorphosis.

Range: 1082 North American species. **Size:** 12-80 mm (1/2-3 1/8").

62. Grasshopper adult

65. Tree cricket

63. Grasshopper nymph

64. Katydid

67. Grasshopper adult

66. Field cricket

ORDER MANTODEA
Mantids

68. European mantid

There are 20 species of praying mantids in North America. They are easily recognized by their elongated prothorax and **raptorial** forelegs. They tend to occupy fairly small territories where they lie in wait for prey. Young nymphs capture very small prey such as fruit flies and aphids, while adults feed on larger prey such as bumble bees, butterflies, wasps and even the occasional hummingbird!

Identifiers: Mobile, triangular head; elongated prothorax; forelegs lined with spikes for capturing prey.

Life Cycle: Eggs are laid in foamy cases containing 50 to 250 eggs depending on species and individual health. Egg cases overwinter in the north.

Range: 20 North American species. **Size:** 10-150 mm (3/8-5 7/8").

PRED

69. Carolina mantid

Chinese mantids

70. 1st instar nymph

71. Adult

ORDER PHASMATODEA
Family Phasmidae
Walkingsticks

PASS
HERB

These herbivores mimic sticks and twigs. They are often found on milkweed growing near the trees and shrubs on which they feed.

72.

Identifiers: Extremely elongate thorax and abdomen; usually wingless, except for one species in south Florida.

Life Cycle: Females drop eggs singly. Eggs overwinter in leaf litter and hatch in the spring. Simple metamorphosis.

Range: 8 North American species.   **Size:** 75-150 mm (3-5 7/8").

ORDER ISOPTERA   Family Rhinotermitidae
Subterranean Termites
Reticulitermes spp.

DEC/S
PASS

Winged reproductive adults from a nearby colony swarm are the only members of this group of social insects that will show up on milkweed. Although they are passersby in the milkweed community, they play an important role as decomposers vital to the prairie and other ecosystems. Through a special symbiosis with protozoans and bacteria, which live in their hind guts, they are able to break down cellulose and return the nutrients to the soil.

Identifiers: Small, soft-bodied, reproductive adults with two relatively long pairs of wings of similar size held flat over the abdomen at rest.

Life Cycle: Termites go through simple metamorphosis in highly organized social colonies in the ground or in wood.

Range: 44 North American species.   **Size:** 5-25 mm (1/4-1").

ORDER DERMAPTERA
Earwigs

These nocturnal insects are frequently seen tucked between milkweed leaves during the day. They eat vegetation, nectar and detritus. Some species prey on caterpillars and other insects' eggs, larvae and pupae.

Identifiers: Pincer-like **cerci**; elongate, flattened insects with short elytra.

73. Earwig

HERB
DEC/S
NECT
PRED

Life Cycle: Females dig cup-shaped nests in the soil and overwinter with their eggs. They stay with the young nymphs for a few days after they hatch. Simple metamorphosis.

Range: 18 North American species.   **Size:** 6-35 mm (1/4-1 3/8").

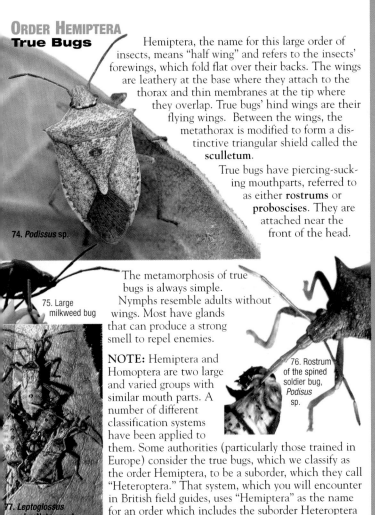

ORDER HEMIPTERA
True Bugs

Hemiptera, the name for this large order of insects, means "half wing" and refers to the insects' forewings, which fold flat over their backs. The wings are leathery at the base where they attach to the thorax and thin membranes at the tip where they overlap. True bugs' hind wings are their flying wings. Between the wings, the metathorax is modified to form a distinctive triangular shield called the **sculletum**.

True bugs have piercing-sucking mouthparts, referred to as either **rostrums** or **proboscises**. They are attached near the front of the head.

74. *Podissus* sp.

The metamorphosis of true bugs is always simple. Nymphs resemble adults without wings. Most have glands that can produce a strong smell to repel enemies.

75. Large milkweed bug

76. Rostrum of the spined soldier bug, *Podisus* sp.

NOTE: Hemiptera and Homoptera are two large and varied groups with similar mouth parts. A number of different classification systems have been applied to them. Some authorities (particularly those trained in Europe) consider the true bugs, which we classify as the order Hemiptera, to be a suborder, which they call "Heteroptera." That system, which you will encounter in British field guides, uses "Hemiptera" as the name for an order which includes the suborder Heteroptera and the two suborders of the order Homoptera, Auchenorrhyncha and Sternorryncha.

77. *Leptoglossus* nymphs. Note scent glands on back.

31

(HERB)

**78.
Neurocolpus
nubilus**

79. and 80. Unidentified miridae

Family Miridae
Plant Bugs

This is by far the largest family of true bugs. They are small—most are less than 9 mm (3/8")—and have soft bodies. They can have a wide variety of colors and patterns, and can be quite abundant. Nearly all are herbivorous.

Identifiers: Bodies seem "squished" at the back end.

Life Cycle: Females insert eggs into plant tissue with a blade-like ovipositor. Most adults live on plants and some are considered crop pests.

Range: 1777 North American species.

**81. Tarnished plant bug,
Lygus lineolaris**

Size: 4-9 mm (1/8-3/8").

• • • • • • • • • •

Family Phymatidae
Ambush Bugs

Ambush bugs are common predators usually found on flowers, where they lie in wait for prey. They are small, but able to capture insects much larger than themselves. They do not bite humans.

82. Ambush bug

**83.
Phymata sp.**

Identifiers: Front legs greatly thickened to accommodate muscles to capture prey; rear end of the abdomen wider than wings.

Life cycle: Ambush bug eggs are glued to plants. Nymphs attack small insects.

Range: 22 North American species.

Size: 8-12 mm (3/8-1/2").

• • • • • • • • • •

Family Reduviidae
Assassin Bugs

Assassin bugs are aggressive predators. They stab their victims, paralyzing them and dissolving their tissues with saliva injected through the wound. They frequently have powerful muscles in their forelegs for grasping and subduing prey. The **rostrum** (beak) is short, curved and fits into a groove between the forelegs. Assassins' body shapes are quite variable.

84. Assassin bug, *Zelus* sp., with monarch caterpillar

Some, like the wheel bug, *Arilus cristatus*, have strange spikes and body expansions. (*See photos below.*)

Identifiers: Elongated head with narrow neck; stout, three segmented, curved rostrum; abdomen in some species widened in the middle so that it extends beyond the wings.

85. Wheel bug attacking monarch chrysalis

Life Cycle: Clutches of up to 50 eggs, laid in cracks in stems or bark or glued to foliage, are sometimes guarded by males. Assassin nymphs are aggressive predators.

Range: 160 North American species.

Size: 12-36 mm (1/2-1 3/8").

• • • • • • • • • •

86. Wheel bug, *Arilus cristatus*

PRED

PRED

87. Damsel bug, *Nabicula subcoleoptrata*

Family Nabidae
Damsel Bugs

Damsel bugs are common predators on low vegetation.

Identifiers: Slender, four-segmented rostrum (beak); conspicuous eyes; front legs slightly thickened; most are yellowish brown but some are shiny black; forewings, when developed, have a number of small cells along the margin.

Life Cycle: Eggs are inserted into plant tissue by females. Nymphs develop over 1-2 months.

Range: 48 North American species.   **Size:** 8-12 mm (3/8-1/2").

• • • • • • • • • • •

Family Tingidae
Lace Bugs

Lace bugs are small, flat insects with lacy patterns on their heads, thorax and wings. Although they feed on the foliage of trees and shrubs, adults can be quite common in milkweed patches near wooded areas.

Identifiers: Flattened bug with lace patterns on head, thorax and wings.

Life Cycle: Females cut slits in leaves to lay eggs. Nymphs are covered with spines and much darker than adults.

88. Lace bug

Range: 157 North American species.   **Size:** 3-6 mm (1/8-1/4").

• • • • • • • • • • •

Family Lygaeidae
Seed Bugs

This is the second largest family of true bugs in North America. There is a great deal of variation in size, shape and color. Most feed on seeds or suck sap, but a few can be predators or scavengers. This family includes the milkweed bugs often found in large groups feeding on milkweed.

Range: 295 North American species.

89. Milkweed bug nymphs congregating on a milkweed pod

Milkweed Bugs

There are two common species of milkweed bugs in North America, the small milkweed bug, *Lygaeus kalmii*, and the large milkweed bug, *Oncopeltus fasciatus*. Their

90. Small milkweed bug, *Lygaeus kalmii*

91. Large milkweed bug, *Oncopeltus fasciatus*

impacts in the milkweed patch can be very different. In spite of their names, these two species are best distinguished from one another by the markings on their wings. The markings on *L. kalmii* form a red X, and on *O. fasciatus* they form a black band. Milkweed bugs use the same orange and black warning colors that monarch and queen butterflies use. Their tendency to congregate probably emphasizes the warning message.

Milkweed bugs feed by piercing food with their long rostrums and injecting salivary enzymes to digest it. They suck up the resulting fluid.

93. Large milkweed bug, *Oncopeltus fasciatus*

92. At first glance, early instar milkweed bug nymphs look like red aphids.

Large Milkweed Bug
Oncopeltus fasciatus

MW
HERB
NECT

Large milkweed bug adults eat milkweed plant matter, mature and maturing milkweed seeds and nectar from milkweed and other flowers. They have been successfully bred to eat sunflower seeds in captive populations.

Adults overwinter in leaf litter, however they cannot survive extreme northern winters. Many of them migrate in late summer and early fall, spending the winter in the central United States. In early summer, some adults migrate back to northern areas.

Identifiers: Wings red or orange with a black band across the middle.

Range: East of the Rockies.

Size: 10-15 mm (3/8-5/8").

CLASS INSECTA

35

In addition to the widespread O. *fasciatus*, there are six other members of the genus *Oncopeltus* whose ranges are restricted to California, Arizona, Texas and/or Florida. The six-spotted milkweed bug, O. *sexmaculatus*, can be found in parts of Florida and Texas.

94. Six-spotted milkweed bug, *Oncopeltus sexmaculatus*

Small Milkweed Bug
Lygaeus kalmii

A study of L. *kalmii* in California has established that they have adapted to eat plants other than milkweed.[3] They were observed acting as scavengers and predators as well, especially in the spring when milkweed seeds, their preferred food, are scarce or non-existent. L. *kalmii*'s ability to cope with milkweed toxins enables them to prey on other milkweed specific organisms. They have been documented eating each other and O. *fasciatus*, as well as monarch chrysalides and small caterpillars.

MW HERB PRED DEC/S

95. (*above*) and 96. (*below*) Note the variation in the width of the red markings on these 2 specimens of small milkweed bug, L. *kalmii*.

Identifiers: Red band across prothorax; markings on wings form a red X.

Range: Throughout US. **Size:** 10-13 mm (3/8-1/2").

97. L. *kalmii* mating

• • • • • • • • • • • •

Family Coreidae
Leaf-footed Bugs

HERB

Most leaf-footed bugs suck juices from plants. They secrete a foul-smelling fluid when disturbed.

Identifiers: Dark colored; head small compared to pronotum; hind tibia of some species dilated and leaf-like.

Life Cycle: Eggs are laid on vegetation. Coreidae overwinter as adults.

Range: 80 North American species.

Size: 9-23 mm (3/8-7/8").

97. *Acanthocephala femorata*

• • • • • • • • • • • •

[3] Root, Richard B. 1986. The life of a Californian population of the facultative milkweed bug, *Lygaeus kalmii*, (Heteroptera: Lygaeidae.), **Proc. Ent. Soc. Wash.**, 88(2), pp. 201-214.

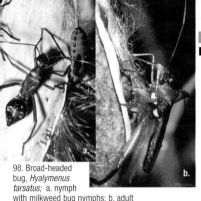

98. Broad-headed bug, *Hyalymenus tarsatus*; a. nymph with milkweed bug nymphs; b. adult

Family Alydidae
Broad-headed Bugs

(HERB)

This is a small family of plant- and seed-eaters. The nymphs of some species mimic ants both in appearance and behavior.

Identifiers: Slender; elongate; head nearly as wide as pronotum; well-developed scent glands.

Life Cycle: Eggs are laid in soil and leaf litter.

Range: 29 North American species.

Size: 10-18 mm (3/8-3/4").

• • • • • • • • • • • • •

Family Pentatomidae
Stink Bugs

These insects discharge foul smelling fluid when they are disturbed. Although most are herbivorous, a few predatory species prey on monarch caterpillars.
Range: 247 North American species.

99. *Eustochistus* sp.

(HERB)

Green Stink Bug
Acrosternum hilare

This bug feeds on plant juices, flowers and fruit from a variety of plants.

Identifiers: Bright green; shield-shaped.

Life Cycle: Keg-shaped eggs are attached to the underside of foliage.

Range: All of North America.

Size: 13-18 mm (1/2-3/4").

100. Green stink bug

101. Green stink bug eggs

Family Pentatomidae
Stink Bugs (*continued*)

Spined Soldier Bug
Podisus spp.

Spined soldier bugs feed on caterpillars, grubs of leaf beetles, and other insects, including each other.

Identifiers: Shield-shaped; pale-brown to yellowish tan; sharp spine on either side of thorax.

Life Cycle: Females lay as many as 40 clutches of 20-30 metallic bronze eggs. Nymphs stay together and feed on plant matter until their first molt, when they become predaceous.

Range: All of North America.

Size: 10-12 mm (3/8-1/2").

102. Spined soldier bug, *Podisus* sp., with milkweed tussock moth caterpillar

HERB
PRED

103. Spined soldier bug, *Podisus* sp., with monarch caterpillar

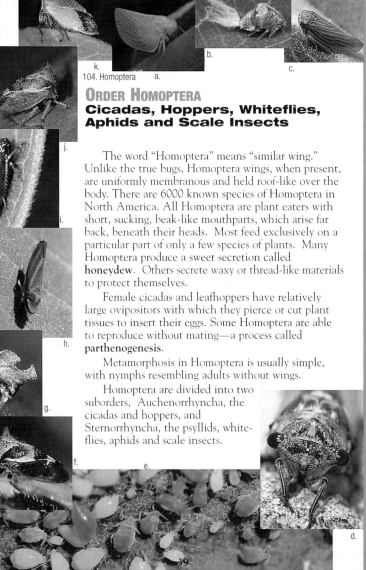

k.
104. Homoptera a.

b.

c.

ORDER HOMOPTERA
Cicadas, Hoppers, Whiteflies, Aphids and Scale Insects

The word "Homoptera" means "similar wing." Unlike the true bugs, Homoptera wings, when present, are uniformly membranous and held roof-like over the body. There are 6000 known species of Homoptera in North America. All Homoptera are plant eaters with short, sucking, beak-like mouthparts, which arise far back, beneath their heads. Most feed exclusively on a particular part of only a few species of plants. Many Homoptera produce a sweet secretion called **honeydew**. Others secrete waxy or thread-like materials to protect themselves.

Female cicadas and leafhoppers have relatively large ovipositors with which they pierce or cut plant tissues to insert their eggs. Some Homoptera are able to reproduce without mating—a process called **parthenogenesis**.

Metamorphosis in Homoptera is usually simple, with nymphs resembling adults without wings.

Homoptera are divided into two suborders, Auchenorrhyncha, the cicadas and hoppers, and Sternorrhyncha, the psyllids, whiteflies, aphids and scale insects.

j.

i.

h.

g.

f. e.

d.

SUBORDER AUCHENORRHYNCHA
Cicadas and Hoppers

Members of this suborder, including the cicadas, treehoppers, froghoppers, leafhoppers and planthoppers, have short, bristle-like antennae and are relatively active. They can be frequent visitors to milkweed, but participate in the milkweed community primarily as prey.

PASS
HERB

105. Cicada

Family Cicadidae
Cicadas

Cicadas are the largest Homoptera in the United States. The nymphs feed underground for an extended period of time. When they come to the surface for their final molt, they climb any available vertical surface and often show up on milkweed plants growing near the trees on which they fed. The adults leave behind a distinctive exoskeleton when they emerge.

106. Cicada nymph to adult metamorphc

The trilling drone of cicadas is a familiar sound of summer throughout much of North America. Cicadas are sometimes called locusts, a name that is more correctly applied to migratory grasshoppers. (*See page 13 for more information.*)

Identifiers: Large, thick insects with membranous wings.

Life Cycle: Nymphs spend 4-17 years feeding on tree roots underground. Winged adults emerge from the mature grub bodies and fly to treetops to mate. Females lay their eggs in twigs, which die off and fall to the ground. When they hatch the nymphs burrow into the ground.

Range: 166 North American species.   **Size:** 25-60 mm (1-2 3/8").

• • • • • • • • • •

Family Membracidae
Treehoppers

PASS
HERB

107.

Treehoppers are small, jumping insects with enlarged **pronotums** (plate-like area on the front and top of the prothorax) resembling thorns. Those found on milkweed are most likely associated with nearby host trees or shrubs. Nymphs do not have enlarged pronotums and are sometimes tended by ants attracted to the honeydew they produce.

Identifiers: Enlarged pronotums, often resembling thorns.

Life Cycle: Treehoppers overwinter as eggs laid on or near host trees. Some females guard their egg masses and young nymphs from predators and parasitic wasps.

Range: 258 North American species.   **Size:** 5-15 mm (1/4-5/8").

• • • • • • • • • •

108.

Family Cercopidae
Froghoppers or Spittlebugs

Adult spittlebugs, often called froghoppers, can be very common on milkweed plants if their preferred host plants, such as goldenrod and clover, are nearby. They serve as prey within this community.

Identifiers: Adults wider to the back, resembling tiny frogs; nymphs encased in a frothy, spittle-like mass.

Life Cycle: Cercopidae overwinter as eggs laid at the angle between leaf and stem of a host plant. Nymphs cover themselves with froth.

Range: 54 North American species.   **Size:** 4-13 mm (1/8-1/2").

Family Cicadellidae
Leafhoppers

Leafhoppers are common, often brightly colored, jumping insects. Each species requires a specific host plant. None are known to feed on milkweed, but many different species will land briefly on milkweed leaves. They are prey within this community.

Identifiers: Similar to froghoppers but more elongate; parallel-sided or tapering to the back; one or more rows of small spines extending the length of the hind tibiae.

Life Cycle: Eggs are usually laid in the soft tissue of host plants. Adults of some species overwinter.

The row of spines, visible on the hind tibiae of the *Grapholocephala* above and the leafhopper in the lower left hand corner, helps identify them as members of the Cicadellidae family.

13. Blue sharpshooter, *Hordnia circellata*

Many secrete honeydew, which attracts ants, flies and wasps.

Range: 2507 North American species.

Size: 2-15 mm (1/16-3/8").

112.
Sharpshooter

114. Planthoppers

SUPERFAMILY FULGOROIDEA
Planthoppers

These wedge-shaped, jumping insects are common, but not as abundant as the other hoppers. Individuals in the same species may be long-winged, short-winged or wingless. There are 11 families of planthoppers identified in North America.

115. Planthoppers feeding on milkweed

a. Silken enclosure and molted exoskeletons

Planthoppers feed on the sap of grasses and herbaceous plants.

Identifiers: Distinct angle between the front and lateral surfaces of the head.

Life Cycle: Simple metamorphosis.

Range: 11 families and over 500 North American species.

Size: 8-10 mm (1/8-3/8").

The planthoppers shown to the right are members of the family Flatidae. Flatidae can be distinguished from other planthoppers by the many small veins on the edges of their wings. They are usually green or brown with wings held nearly vertical at rest. The nymphs are protected by long threads of wax produced from the surface of their bodies. There are 58 North American species of Flatidae.

These planthoppers were observed feeding exclusively on milkweed in southwestern Pennsylvania. The nymphs fed in groups of three to six individuals and began emerging as adults in late July.

116. Along with the winged and unwinged green aphids and their shed nymphal skins, this photo displays a sampling of the diverse community of predators and parasites that depend on them. Aphids are frequently prey for ladybug larvae (2 in upper left), syrphid fly larvae (1 in lower right) as well as being hosts for parasitic wasp larvae. The brown aphids in the left center are the empty skins of "aphid mummies"— aphids parasitized by wasp larvae. Note the round hole cut by the emerging adult wasps at the back of the aphid exoskeleton.

SUBORDER STERNORRHYNCHA
Psyllids, Whiteflies, Aphids and Scale Insects

These are usually inactive Homoptera with long, thread-like antennae.

Family Aphididae
Aphids

Aphids have piercing-sucking mouthparts through which they draw plant juices. They are pear shaped, soft-bodied insects with a pair of wax secreting tubes called **cornicles** on the back of their abdomens.

In addition to the oleander aphids described below, green and black aphids are common on milkweed. Although no species of ant is known to tend oleander aphids, other aphids are often tended by ants. Aphids secrete a sweet substance called honeydew, which ants ingest and take back to their colonies. In return,

117. Ants often tend aphids, collecting the honeydew they produce.

118. Brown and black aphids above are parasitized "aphid mummies."

the ants protect the aphids from predators.

Aphids have a complex life cycle which allows them to multiply rapidly. Overwintering eggs hatch in the spring as wingless females. These females give birth **parthenogenetically**—without mating—to more wingless females. Winged generations of females are produced periodically throughout the summer and fly to other plants to begin new colonies, which continue to reproduce parthenogenetically. In many species a last generation of winged females in the fall produces a generation of wingless males and egg laying females, which mate to produce overwintering eggs.

Ladybug beetles, lacewings and some syrphid fly larvae consume aphids. Tiny braconid and chalcidoid wasps parasitize them.

Range: 1351 North American species.

119. Braconid wasp, *Lysiphlebus testaceipes*, emerging from aphid

Oleander Aphids
Aphis nerii

Yellow oleander aphids are quite common on milkweed. *Aphis nerii* is an exotic species that probably originated in the Mediterranean region. Its primary host is oleander of the plant family Apocynaceae—a family closely related to Asclepiadaceae.

MW

HERB

In this country, the oleander aphid appears to be an **obligate parthenogenetic** species—a species which always reproduces asexually. Males do not occur in the wild. Winged adult females develop in overcrowded conditions and fly to colonize other plants.

Oleander aphids can be parasitized by wasp larvae from the species *Lysiphlebus testaceipes* and *Aphidis colemani*. Once matured, the wasps emerge, leaving the hollow shell of the aphid behind. The parasitized aphids are coppery or black in color and are referred to as aphid mummies. (*See photos above.*)

Identifiers: Soft, yellow, pear-shaped bodies.

Life Cycle: Simple metamorphosis. Nymphs are **parthenogenetically** produced and born live.

Range: Worldwide.

Size: 1.5-2.6 mm (1/16-1/8").

120. *Aphis nerii* clustered around vein on underside of a milkweed leaf.

121. Lacewing head close-up

122. Lacewing eggs

123. Lacewing larva

124. Lacewing larva with aphid

ORDER NEUROPTERA
Net-veined Insects

The word "Neuroptera" means "nerve wing" and refers to these insects' elongate-oval, transparent wings with many veins. Four similarly sized wings are usually held roof-like over the body at rest. In flight they beat in a seemingly uncoordinated fashion. Neuroptera undergo complete metamorphosis. Larvae are mostly predaceous with long sickle-shaped mandibles and do not resemble adults.

Family Chrysopidae
Green Lacewings

Lacewings are common insects. Their larvae, sometimes called aphid lions, prey chiefly on aphids, but also eat small caterpillars and other small, soft-bodied insects.

Identifiers: Pale green adults, with prominent, gold or coppery eyes; thread-like antennae 2/3 body length; clear, green-veined wings at least 1/4 longer than body.

Life Cycle: Eggs are laid suspended from leaves by slender silken stalks. Nymphs are predaceous.

Range: 87 North American species.

Size: 10-15mm (3/8-3/4").

Pred

125. Green lacewing

126. Swamp milkweed leaf beetle

127. Grapevine beetle

128. Japanese beetle

130. Soldier beetle head

131. Pennsylvania leatherwing

137. *Trirhabda virgata*

136. Click beetle

135. **Milkweed beetle**

134. Firefly

133. Seven spot ladybug

132. Ladybug

ORDER COLEOPTERA
Beetles

Coleoptera—the beetles—is the largest order in the animal kingdom, including one third of all known insects. They are found in almost every type of habitat that supports insects. The name "Coleoptera" means "sheath wing" and refers to the tough, armor-like forewings (**elytra**) that meet in a straight line down the middle of their backs. The elytra cover their membranous hind wings, which are used for flying. Beetles have prominent compound eyes. Their chewing and biting mouthparts, with well-developed mandibles, enable them to eat a wide range of materials. Beetles can be predators, herbivores, scavengers and parasites. Some eat leaves, others bark, dung, or fabric. Most adult beetles can fly, but do so only for short distances. Beetles go through complete metamorphosis. Most species have one generation a year. Their larvae, called **grubs**, can be predaceous or herbivorous.

46

138. Fiery caterpillar hunter,
Calosoma scrutator

Family Carabidae
Ground Beetles

This is a large family of predaceous beetles. Most are nocturnal. They move rapidly and may climb up trees and vegetation in search of prey, including caterpillars. A few eat pollen, berries and seeds. Many give off an unpleasant odor when handled.

Species of the genus *Calosoma* are called caterpillar hunters and are often abundant in fields.

Identifiers: Conspicuous prothorax; narrow head; long legs with spurs on the tibia; most with grooved (**striate**) elytra; thread-like antennae arise between large compound eyes.

Life Cycle: Adults overwinter and live 2-3 years. Predaceous larvae take a year to grow from eggs to adults.

Range: More than 2220 North American species.

Size: 3-36 mm (1/8-1 3/8").

139. Adult, larva and pupa of the caterpillar hunter beetle, *Calosoma calidium*.

Family Cantharidae
Soldier Beetles

Soldier beetles are common on milkweed flowers where they take nectar and pollen. Both larvae and adults prey on aphids and other soft-bodied, small insects.

Identifiers: Protruding heads; soft elytra usually yellow, red or orange and covered with short downy hairs; resemble lightening bugs without light producing organs.

Life Cycle: Eggs are laid in soil. Larvae pupate in cells in soil.

Range: 468 North American species.

Size: 9-15 mm (3/8-5/8").

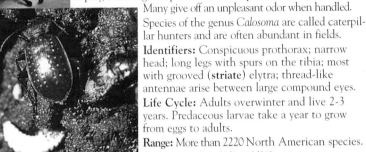

140. Pennsylvania leatherwing, *Chauliognathus pennsylvanicus*, searches a milkweed flower for pollen and nectar (*above left*). 141. Close up of head (*above right*). 142. Another unidentified soldier beetle (*below left*).

143. Head and pronotum *(left)*
144. Whole body of Pennsylvania firefly,
Photuris pennsylvanicus *(right)*

Family Lampyidae
Fireflies

PASS
PRED

These familiar, soft-bodied insects can often be found resting on milkweed.

Identifiers: Most have light producing organs; pronotum fully or partially conceals head from above.

Life Cycle: Firefly larvae feed on small animals, like snails, in debris. The food of adults is unknown. It is believed that some adults don't eat, however, some (*like the one to the right*) seem to be attracted to nectar.

Range: 136 North American species.

Size: 5-20 mm (1/4-3/4").

145. *Pyractonema* sp. (*lower photo*)

Family Elateridae
Click Beetles

HERB

This is a large group of very common beetles with a distinctive, elongated and somewhat flattened shape. They get their name from the sharp clicking sound the adults make as they turn themselves upright when they are turned over on their backs. Some adults do not feed, and those that do are herbivores.

Identifiers: Elongate, flat, and brown or black; large, flexible prothorax with pointed rear corners.

Life Cycle: Click beetle larvae, called wireworms, live in the soil and feed on roots and germinating seeds.

Range: 800 North American species.

Size: 2-50 mm (1/16-2").

146. Click beetles, like this one, are often found nestled in the new growth at the top of *A. syriaca* plants.

Family Nitidulidae
Sap Beetle

HERB

These small, dark, robust beetles with clubbed antennae are herbivores which can be found in milkweed flowers. The most conspicuous is the red- and black-marked *Glischrochilus*.

Identifiers: Abruptly clubbed antennae; variable shapes; many are robust.

Range: 215 North American species.

Size: 1.5-12 mm (1/32-1/2").

147. *Glischrochilus* sp.

148. Multicolored Asian lady beetle, _Harmonia axyridis_

149. Seven spotted ladybug, _Coccinella septempunctata_

PRED

Family Coccinellidae
Ladybug Beetles

Nearly everyone recognizes the classic red ladybug with her characteristic black spots, but there is a great variation in the size, number of spots and patterns in the 400 coccinellid species found in North America. Ladybugs can be the most abundant predator in a milkweed community. Both the larvae and adults prey on aphids and small caterpillars. Some species specialize on mealybugs, mites or scale insects. One species is herbivorous. A recently introduced species, the multicolored Asian lady beetle, _Harmonia axyridis_, has made itself unpopular by congregating to overwinter in homes.

Identifiers: Oval to nearly spherical; strongly convex on top and nearly flat on the bottom; head partly or completely hidden below the pronotum; background color often bright red, orange or yellow, but can be black; antennae short and clubbed.

Life Cycle: Bright yellow eggs are laid on leaves near prey species. Most larvae are active predators. They are spindle-shaped with spines, spots and bands—resembling tiny alligators. Some produce waxy filaments. Adults overwinter and many species form large congregations.

Range: 400 species in North America, 5000 species worldwide.

Size: 0.8-10 mm (1/16-3/8").

158. Convergent ladybugs, _Hippodamia convergens_, mating

157. Eggs

156. Larva

Shed exoskeleton

154. Larva molting

153. Larvae, genus _Scymnus_

152. Larva eating aphid

150. Congregation of convergent ladybugs in southeastern Arizona

151. Pupa

49

Gallery of Ladybug Beetles

The ladybug beetles below are just a small sampling of the
species that are active participants in a milkweed community.

159. Undulated ladybug
Hyperaspis undulata
3 mm (1/8")

160. *Hyperaspis binotata*
3 mm (1/8")

161. *Scymnus* sp.
1.5 mm (1/16")

162. Spotted ladybug
Coleomegilla sp.
5-8 mm (3/16-5/16")

163. *Brachyacantha
ursina*
3-4 mm. (1/8")

164. Convergent ladybug
Hippodamia convergens
3-6 mm (1/8-1/4")

165. Multicolored Asian lady
beetle, *Harmonia axyridis*
6 mm (1/4")

166. Eastern ladybug
Coccinella transversoguttata
5 mm (3/16")

167. Spotless ladybug
Cycloneda sanguinea
3-5 mm (1/8-3/16")

Family Meloidae
Blister Beetles

Blister beetles are common on flowers and foliage. When disturbed they exude blood which can cause blistering. Adults are herbivores. Larvae are parasites that usually feed on grasshopper eggs, although some feed on the eggs and larvae of bees.

168. *Epicauta* sp.

Identifiers: Elongate and slender; usually black or brown; pronotum narrower than wings.

Life Cycle: Larvae go through **hypermetamorphosis**—the larval instars are quite different in form. The first instar is long-legged and active. They climb onto flowers and attach themselves to bees visiting the flowers. The bees carry them back to their nests. Subsequent instars are grub-like or maggot-like and attack bee eggs.

Range: 335 North American species.

Size: 9-28 mm (3/8-1 1/8").

• • • • • • • • • •

169. Tumbling
flower beetle

Family Mordellidae
Tumbling Flower Beetles

These beetles are common on flowers. They are difficult to catch as they tumble off the flowers and take flight when disturbed.

Identifiers: Humpbacked; head bent down; pointed abdomen extending beyond the forewings.

Life Cycle: Larvae occur in rotten wood or plant pith. Some are leaf and stem miners.

Range: 204 North American species.

170. Tumbling
flower beetle

Size: 3-7 mm (1/8-3/8").

• • • • • • • • • •

Family Scarabaeidae
Scarab Beetles

This is one of the largest families of beetles. Japanese beetles, *Popilla japonica*, in eastern North America; the bumble flower beetle, *Euphoria inda*, in eastern North America; the flower chafer,

180. Flower chafer,
Euphoria sepulcralis

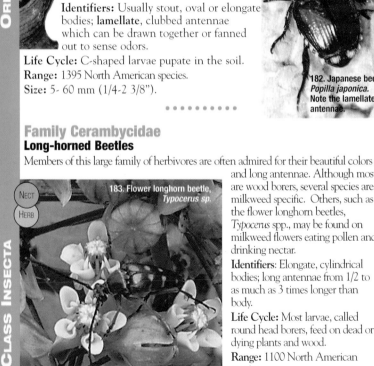

Nect
Herb

181. Rose chafer,
Macrodactylus subspinosus

Euphoria sepulcralis, in southeastern to midwestern US; and the rose chafer, *Macrodactylus subspinosus*, are among the species of scarab beetles that may be found on milkweed. The adults drink nectar and eat pollen and plant tissue.

Identifiers: Usually stout, oval or elongate bodies; **lamellate**, clubbed antennae which can be drawn together or fanned out to sense odors.

Life Cycle: C-shaped larvae pupate in the soil.

Range: 1395 North American species.

Size: 5- 60 mm (1/4-2 3/8").

182. Japanese beetle,
Popilla japonica.
Note the lamellate
antennae.

• • • • • • • • • •

Family Cerambycidae
Long-horned Beetles

Members of this large family of herbivores are often admired for their beautiful colors and long antennae. Although most are wood borers, several species are milkweed specific. Others, such as the flower longhorn beetles, *Typocerus* spp., may be found on milkweed flowers eating pollen and drinking nectar.

Nect
Herb

183. Flower longhorn beetle,
Typocerus sp.

Identifiers: Elongate, cylindrical bodies; long antennae from 1/2 to as much as 3 times longer than body.

Life Cycle: Most larvae, called round head borers, feed on dead or dying plants and wood.

Range: 1100 North American species.

Size: 6-75 mm (1/4-3").

Milkweed Longhorn Beetles
Tetraopes spp.

There are eight species of this genus which use milkweed as a host plant. These species are distinguishable largely by the number and arrangement of spots on their wings. The beetles make squeaking noises when disturbed by rubbing together rough spots on their thorax. They leave characteristic chew holes on the tips of milkweed leaves. (*See page 12 for more information.*)

MW
HERB

184. *Tetraopes tetraophthalmus*

Identifiers: Elongate, nearly cylindrical; red elytra with black spots; *T. tetraophthalmus*, gray antennae.

Life Cycle: Eggs are laid on milkweed stems near the ground. The larvae burrow into and eat the stems and roots of milkweed. They overwinter underground and pupate in the spring.

Range: Two most prevalent: *T. tetraophthalmus*, eastern US and Canada; *T. femoratus*, all of North America except the Atlantic coast.

Size: 10-13 mm (3/8-1/2").

185. *Tetraopes femoratus*

● ● ● ● ● ● ● ● ● ● ●

Family Chrysomelidae
Leaf Beetles

This is a large family with many common species. Many have bright metallic colors. Adults feed on leaves and flowers, while larvae feed on roots and leaves or tunnel through them. Some, like the spotted cucumber beetle and the Colorado potato beetle, are serious crop pests. The milkweed leaf beetle, *Chysochus* sp., and the swamp milkweed

186. Leaf-mining leaf beetle, subfamily Hispinae

188. Note the enlarged hind femor adapted for jumping on this chrysomelid.

187. Spotted cucumber beetle, *Diabrotica undecimpunctata howardii*

53

leaf beetle, *Labidomera clivicolis*, from this family feed on milkweed.

The bright colored tortoise beetles with clear elytra are part of this family. The argus tortoise beetle, sometimes called the milkweed tortoise beetle, *Chelymorpha cassidea*, is often found on milkweed, however, it does not feed on it. It is most likely feeding on bindweed or wild morning glory common in many midwestern milkweed patches.

191. *Trirhabda virgata*

Identifiers: Short antennae, less than 1/2 the length of the body; body oval, usually convex; some with bright metallic colors; some with enlarged hind femora for jumping.

Life Cycle: Larvae feed on foliage and roots; some, from the subfamily Hispinae, are leaf miners. Many pupate in the soil. Adults eat leaves and flowers.

Range: 1474 North American species.

Size: 1-13 mm (1/16-1/2").

192. Bean leaf beetle, *Cerotoma trifurcata*

189. Argus tortoise beetle, *Chelymorpha cassidea*, larva, pupa and adult. Note the frass carrying behavior of the larva *(above and to the left)*.

190.Tortoise beetle, *Deloyala guttata (below)*

194. *Labidomera clivicolli* larva *(below)*

Swamp Milkweed Leaf Beetle

Labidomera clivicollis

These beetles can feed on common milkweed but are more abundant on swamp milkweed. Kathryn Eickwort found that generalized arthropod predators, including damsel bugs, predatory stink bugs and syrphid fly larvae, were important predators of swamp milkweed beetles.[4] Those predators that climbed onto

193. *Labidomera clivicollis* adult

[4] Eickwort, K. R. 1977. Population dynamics of a relatively rare species of milkweed beetle (Labidomera). *Ecology*, Vol. 58, No. 3: 527-538.

milkweed plants from the ground accounted for 54% of the deaths of milkweed leaf beetle larvae. This suggests that the beetles' preference for laying eggs on swamp milkweed may be related to the wet conditions in which those plants grow. Standing water may offer relatively inactive swamp milkweed leaf beetle larvae protection from some ground dispersed predators.

195. *Labidomera clivicollis* larvae

Identifiers: Head and pronotum black; elytra red with black marks across midline forming an "X"; both larvae and adults are dome shaped.

Life Cycle: Elongate yellow eggs are laid on milkweed leaves. The larvae drop to the ground to pupate. Adults emerge by late summer. These beetles often overwinter as adults in the shrivelled, wooly leaves of mullein plants.

Range: North America.

Size: 10-13 mm (3/8-1/2").

Milkweed Leaf Beetle
Chrysochus spp.

Identifiers: Glossy, dark metallic blue/green.

Life Cycle: Yellow eggs are laid on a host plant at or near ground level. Larvae feed on roots and pupate in soil. Adults emerge at the end of summer and feed on milkweed leaves.

Range: Western US and Canada.

Size: 9-10 mm (3/8").

• • • • • • • • • •

196. Milkweed leaf beetle, *Chrysochus* sp.

Family Curculionidae
Weevils

The Family Curculionidae is the largest family of insects with 40,000 species worldwide and 2500 in North America. They are hard-bodied beetles with heads elongated to form a slender down-curved snout. Several species can be found on milkweed.

Milkweed Weevils
Rhyssomatus lineaticollis

This common milkweed herbivore causes characteristic damage on milkweed leaves,

197. Milkweed weevil on leaf

55

chewing holes through the mid-vein and leaving behind a dried mass of latex. Another characteristic sign of weevils is the elongated hole an ovipositing female chews in a milkweed stem.

Identifiers: Small, dark, hard-bodied beetle with elongated snout; elbowed antennae part way down snout.

Life Cycle: The female lays a series of eggs in a hole she bores in the milkweed stem. Larvae feed inside the stem and can be easily observed in the characteristic openings.

199. Milkweed weevil on stem

198. Milkweed weevil damage

Range: From Texas to North Dakota, east to New York and south to Florida.

Size: 1-4.5 mm (1/16-3/16").

ORDER MECOPTERA
Scorpionflies
Family Panorpidae
Common Scorpionflies

PRED
NECT
HERB

Adults feed on slow moving or dead insects, nectar and rotting fruit. Kathleen Eickwort found these unusual looking insects to be a significant predator of milkweed leaf beetle larvae.[4]

Identifiers: Snout-like heads with biting mouthparts; antennae long and thread-like; males with upward curving genitalia resembling a scorpion stinger.

Life Cycle: Larvae are caterpillar-like and scavenge dead insects and other organic matter. Complete metamorphosis.

Range: East of the Rocky Mountains; 394 North American species.

Size: 12-20 mm (1/2-3/4").

200. *Panorpa* sp.

201. White-lined sphinx, *Hyles lineata*

ORDER LEPIDOPTERA
Butterflies and Moths

202. This fritillary butterfly sucks nectar through its proboscis.

The word "Lepidoptera" means "scale wing." The name refers to the overlapping, often colorful, scales that cover these insects' four membranous wings and usually their bodies as well.

Adult Lepidoptera feed principally on nectar and other liquid food. Some moths live only a short time and do not feed. The mouths of adult Lepidoptera that do feed are straw-like **proboscises** for sucking fluids. When not in use, they are usually held coiled beneath the head.

203. This female Promethius moth has no proboscis and will not feed.

All moths and butterflies go through complete metamorphosis, hatching from eggs as caterpillars and pupating before becoming adults. Caterpillars have chewing mouthparts and, with only a few exceptions, eat vegetation. They grow through several **instars**—stages between molts. Caterpillars can be furry or smooth. They usually have well-developed heads; three pairs of true-legs—one pair on each of three thoracic segments; and five pairs of short, fleshy prolegs, equipped with tiny, crochet hooks and located on the middle

57

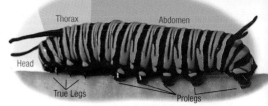

204. Right: Monarch caterpillar body regions and legs.
205. Below: a. Female polyphemus moth with thread-like antennae; b. plumose antennae on a male cecropia moth to assist in picking up the scent of females from a distance; c. sphinx moth spindle-shaped antennae and proboscis; d. skipper proboscis and antennae with hooked ends; e. monarch antennae and proboscis.

four segments and the last segment of their abdomens. Caterpillars produce silk from a modified salivary gland, called a **spinneret**, located under their heads. They use the silk threads to attach themselves to the surface of their host plants, make shelters, spin cocoons or attach their pupal stages to surfaces.

Moths and butterflies differ in their body structure and behavior. Most butterflies are active during the day and most moths are active at night. All butterfly antennae are knobbed at the ends. Moths can have thread-like, clubbed or feathery antennae. Male moths with feathery antennae use the increased surface area to pick up the scent of a female from far away. Caterpillars of some moths create silken cocoons or other structures in which to pupate and some pupate underground. Most butterflies pupate as **chrysalides** (singular chrysalis)—a pupal form without an external shelter.

Most Lepidoptera in the milkweed community are attracted to nectar. With the exception of those adapted to eat milkweed, the impact of the various butterflies and moths in the milkweed community is similar. Many are pollinators for milkweed, many drink nectar and some fall prey to the predators hunting in this busy community.

The moths and butterflies shown here are representative of a few common families. Listing all moths and butterflies that are likely to visit milkweed patches across the continent is beyond the scope of this field guide. We recommend using a butterfly and moth field guide to identify the species in your milkweed patch.

58

206. Pipevine swallowtail, *Battus philenor*

207. Eastern tiger swallowtail, *Pterourus glaucus*

NECT
HERB

Family Papilionidae

Swallowtails

North American swallowtails are large, brightly-colored butterflies with tailed hindwings.

Range: Fewer than 30 North American species.

Size: 54-140 mm (2 1/8-5 1/2").

209. Cabbage white, *Pieris rapae*

NECT
HERB

Family Pieridae
Whites and Sulphurs

208. Clouded sulphur, *Colias philodice*

Pierids are usually some shade of white, yellow or yellowish green.

Range: 55-60 North American species.

Size: 32-51 mm (1 1/4-2").

211.
American copper,
Lycaena phloeas american

210.
Purplish copper,
Epidemia helloides

Family Lycaenidae
Gossamer Wings

This family of small butterflies includes the coppers, blues, hairstreaks and the harvester.

Range: 100 North American species.

Size: 22-51 mm (7/8-2").

NECT
HERB

Edward's hairstreak,
212. *Satyrium edwardsii* 213.

59

Family Nymphalidae
Brush-footed Butterflies

Brush-footed butterflies are medium to large butterflies with reduced forelegs. They include monarchs, admirals, fritillaries, checkerspots, crescents, painted ladies, anglewings, longwings, viceroys and tortoiseshell butterflies. Viceroy butterflies (*below right*) mimic the color and patterns of poisonous monarch butterflies. This may afford them some protection from birds and mammals that have had experience with monarchs. Note the black stripe across the viceroy's hind wings.

Range: 160 North American species.

Size: 38-76 mm (1 1/2-3").

214.
Great spangled fritillary,
Speyeria cybele

216. Red admiral,
Vanessa atalanta

217.
Comma,
Polygonia comma

215. Silvery checkerspot,
Charidryas nycteis

218.
Pearl crescent,
Phyciodes tharos

219. Viceroy,
Basilarchia archippus

NECT
HERB

SUBFAMILY DANAINAE
Milkweed Butterflies

There are 157 species of milkweed butterflies worldwide, but only 4 have been recorded in the United States and Canada. All have reduced forelegs and scaleless antennae. Danaids are known for their strong soaring flight.

Identifiers: North American Danaids: Orange to brown with black veins and markings; reduced forelegs; males with scent patches on the upper side of their hind wings and **hair pencils**—brush-like extensions in their abdomens—which, in some species, are used to waft pheromones that attract females during courtship.

Life Cycle: Milkweed butterflies lay eggs singly on plants within the milkweed family. The caterpillars are striped and have one to three pairs of fleshy tentacles on their bodies. They feed on plants in the milkweed family and **sequester** toxins from the milkweed in their bodies. The bright stripes of the caterpillars and orange or red wings of the adults announce their poisonous nature and unpleasant taste.

220. Monarch,
Danaus plexippus

Range: Worldwide. **Size:** 75-100 mm (3-4").

Monarchs
Danaus plexippus

Monarchs are sturdy butterflies with a 40-55 mm (3-4") wingspan. Male monarchs have a scent patch that appears as a black spot over a vein on their hind wings, although it does not seem to be used in courtship in this species. Monarch caterpillars feed exclusively on plants in the milkweed family.

221. Latex

Monarch egg

Monarchs lay eggs one at a time on milkweed plants, most frequently on the underside of leaves. When milkweed is scarce, they may load a single plant with eggs, but they usually lay only one egg on a plant. In the field it is easy to mistake small drops of hardened milkweed latex for monarch eggs. The eggs have ridges and taper to a point on top. Latex droplets are smooth and round. The black head capsule of the caterpillar can be seen inside eggs about to hatch.

222. Egg

When temperatures are sufficiently warm—between 20-27° C (70s and upper 60s F)—the eggs hatch three to five days after they are laid. In cooler temperatures they can take as many as 20 days to hatch. A newly-hatched caterpillar often eats its eggshell first. It will then eat the milkweed leaf, beginning with the small hairs on the surface, and frequently leaving a characteristic, arched hole in the leaf.

223. 1st instar

The caterpillars eat, grow and molt their outer skins four times, going through five **instars**. The fifth instar caterpillars are about 5 cm (2") long and have yellow, black and white stripes and four fleshy black tentacles—two in front

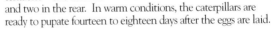

224. Molting

and two in the rear. In warm conditions, the caterpillars are ready to pupate fourteen to eighteen days after the eggs are laid.

225a. 5th instar caterpillar

225b. Monarch metamorphosis from "J" to chrysalis

About ten days after it is formed, the chrysalis begins to darken and the familiar patterns of the monarch butterfly's bright, orange and black wings become visible under a transparent **cuticle**. The butterfly **ecloses,** pumps up and dries its wings and is ready to fly in a matter of hours. Adult monarchs feed on nectar from a wide variety of flowers. Females are sexually mature in four to five days, males in three to five.

226. Monarch
metamorphosis
from chrysalis
to adult

Three or more generations emerge
each summer. Those eclosing in June
and July have an adult life span of four to
five weeks. In the north, the final generation of monarchs
become adults in mid-August through mid-September. Most of these butterflies are reproductively immature—in a state of **reproductive diapause**. They migrate to overwintering sites in Mexico and along the Pacific coast of California, where some survive up to eight months. (*See map below.*)

A small population of reproductively active monarchs persists through the winter in the southern United States and in the lower elevations in the vicinity of the overwintering sites in Mexico.

The exact cues and dynamics of monarch migration are the subject of much research and inquiry. It is known that most monarchs east of the Rocky

FALL

Mountains migrate to the oyamel fir forests, 3000 meters (10,000 feet) above sea level in the mountains of Michoacan in central Mexico. Most west of the Rocky Mountains gather in aggregations near the coast in California. The degree to which these populations mix is unknown. Some individuals may cross the continental divide or stray off the most common paths enough to join the other population,

but recent research suggests that the populations are genetically distinct to some degree.[5]

In clusters of fifty to thousands of individual butterflies, overwintering monarchs wait out the winter in cool

227. Overwintering monarch cluster in Michoacan, Mexico

SPRING

conditions that allow them to remain fairly inactive. In late February or early March the winter clusters disperse. The monarchs break their **diapause** and become reproductive. Those from Mexico arrive in Texas and the southern tier of US states in late March and their offspring continue the journey north, repopulating much of the North American continent over the course of two generations.

Queen
Danaus gilippus

228. Queen butterflies, *Danaus gilippus*

229. Queen caterpillar

MW
HERB
NECT

Queen adults are smaller and darker than monarchs. They have a ruddy brown color, instead of the monarch's orange, and finer black markings on the veins. The black on the upper portion of their wings is restricted to the outer margins. Queen eggs are indistinguishable from monarch eggs, causing some early identification difficulties for monarch observers in the south. The caterpillars have three pairs of tentacles and are dependent on the milkweed family as a host. Male adults must

[5] Altizer, S.M. 2001. Migratory behavior and host-parasite co-evolution in natural populations of monarch butterflies infected with a protozoan parasite. ***Evol. Ecol. Res.***, 3:611-632.

MW
HERB
NECT

230. Queen butterfly, *Danaus gilippus*

drink nectar from certain flowers in the sunflower family to get a chemical needed to produce a pheromone used in courtship and mating.

Queens are common in the southwestern United States, around the Gulf to Florida and into southern Georgia, but they cannot survive cold winters and are not found in northern milkweed communities.

MW
HERB
NECT

Soldier
Danaus eresimus

This butterfly is very similar to *Danaus gilippus* with a lighter, more russet, upper wing surface and a watermark appearance on the lower surface of the hind wing. Larvae are similar to queen larvae, having three pairs of tentacles. Soldiers are found all year round in south Florida. In the fall they become uncommon in south Texas and only occasional in southeast Arizona.

231. Soldier butterfly, *Danaus eresimus*

MW
HERB
NECT

Tiger Mimic-Queen
Lycorea cleobaea

Also called the tropical milkweed butterfly, this butterfly is only rarely seen in south Texas and south Florida. It has long, rounded wings. The caterpillar has only one pair of tentacles located near the head.

• • • • • • • • • •

NECT
HERB

Family Hesperiidae
Skippers

Skippers are not considered true butterflies. They are strong fliers with stout bodies and clubbed

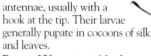

234. Skipper

antennae, usually with a hook at the tip. Their larvae generally pupate in cocoons of silk and leaves.

Range: 250 species in North America.

Size: 13-64 mm (1/2-2 1/2").

232. Long-tailed skipper, *Urbanus proteus*

233. Delaware skipper, *Atrytone delaware*

235. Long dash, *Polites mystic*

• • • • • • • •

Family Sphingidae
Sphinx or Hawk Moths

Sphinx moths are medium to large, heavy-bodied moths. Because of their rapid wing beat, day flying sphinx moths are often mistaken for hummingbirds or large bees.

Range: 124 North American species.

Size: 32-155 mm (1 1/4-6").

236. White-lined Sphinx, *Hyles lineata*

• • • • • • • • • • •

Family Arctiidae
Tiger Moths

Identifiers: Tiger moths are heavy-bodied, small to medium-sized moths. Many are light colored, often with bright spots or bands.

Life cycle: Eggs are laid in flat masses or scattered over vegetation. The caterpillars are usually hairy and many eat toxic substances, like milkweed. They pupate in loose cocoons spun out of silk and their own hairs. Adults generally do not feed.

Range: 264 North American species.

Size: 12-80 mm (1/2-3 1/8").

Milkweed Tiger Moth
Euchaetias egle

This tiger moth is commonly called the milkweed tussock moth. Its eggs are laid in fuzzy clutches on milkweed leaves. Young larvae stay together and often completely defoliate milkweed plants. When disturbed, the caterpillar rolls into a ball and falls to the ground. The bright colors and tufts on the late instar caterpillars have earned them the name "Harlequin Caterpillar." Once the tufts appear, they are often solitary. They overwinter as pupae. The moths are nocturnal and have silver to grayish brown wings. The abdomen is yellow with three rows of black spots. (*See page 12-13 for more information.*)

237. *Euchaetias egle* caterpillars 1st instar (*above*). 238. Last instar (*below*)

239. Milkweed tussock moth, *Euchaetias egle*

MW
HERB
NECT

Two other species of tiger moth caterpillars feed on milkweed.

Unexpected Cycnia
Cycnia inopinatus
These are fairly rare prairie habitat moths.

241c. *Cycnia inopinatus* caterpill

Delicate Cycnia
Cycnia tenera
Also called dogbane tiger moths, these caterpillars feed on milkweed and eat dogbane.

241a. Delicate cycnia moth and b. Delicate cycnia caterpillar

HERB
NECT
Other species of Arctidae nectar on milkweed. The Virginia ctenucha and yellow-collared scape moth from this family are easily confused with the grape leaf skeletonizer, *Harrisina americana*, below.

242. Yellow-collared scape moth, *Cisseps fulvicollis*

HERB
NECT
Family Pterophoridae
Plume Moths
Plume moths are common, slender moths with long legs and narrow wings divided into feathery lobes. They hold their wings at right angles to their bodies when at rest.
Range: 146 North American species. **Size:** 12-40 mm (1/2-1 5/8").

243. Plume moth

HERB
NECT
Family Zygaenidae
Smoky Moths

244. Smoky moth, grape leaf skeletonizer, *Harrisina americana*

Smoky moths are small, black or brown moths with thinly scaled wings.   **Size:** 16-28 mm (5/8-1").

HERB
NECT
Family Pyralidae
Pyralid Moths
Pyralid moths are small, very common moths with large palps which extend beyond their heads like a snout.
Range: 1100 species in North America.
Size: 10-55 mm (3/8-2").

245. Pyralid moth

HERB
NECT
Family Yponomeutidae
Ermine Moths
Ermine moths are small, common moths with spotted or boldly patterned wings. Adults wrap their wings around their bodies at rest.
Size: 12-40 mm (1/2-1 5/8").

246. Ailanthus webworm moth, *Atteva punctella*

ORDER DIPTERA
Flies

The order Diptera includes mosquitoes, midges and flies. They are abundant not only in the number of species, but in the number of individual insects as well. Diptera have a variety of impacts in the milkweed community. Some are predators and parasites, others are nectivores and decomposers. They are potentially pollinators as well,

247. Dipteran trapped in milkweed blossom

but it is not unusual to find a dipteran hanging lifeless from a milkweed blossom, one or more legs trapped in the flower column. (See page 17.)

Many Diptera mimic bees or wasps, and can be difficult to identify. The feature most easily observed, that distinguishes them from all other orders, is their wings. Diptera have one pair of functional wings and a pair of tiny, highly-altered wings, called **halteres**, which are used to stabilize their bodies in flight. Most have large, compound eyes and mouthparts adapted to suck, lap or pierce. Antennae are often short and inconspicuous, but they can be long and thread-like. Diptera are relatively soft-bodied and undergo complete metamorphosis. The soft, legless and headless dipteran larvae are called **maggots**.

248. Thick-headed fly, family Conopidae

There are two suborders of Diptera differing primarily in the structure of the antennae.

SUBORDER NEMATOCERA

Members of the suborder Nematocera are flies with long antennae—usually with more than five segments. They include crane flies, family Tipulidae; mosquitoes, family Culicidae; and midges, family Chironomidae. Most larvae are aquatic or live in moist soil. Because of the abun-

250. Mosquito

249. Midge, *Chironomus attenuatus* dance of some of these families, it is likely that, from time to time, they will be found on milkweed. Their roles

251. Crane fly

in the milkweed community are usually as prey or passerby. An exception is the biting midges, family Ceratopogonidae. Though some suck the blood of mammals and birds, they are mostly parasites or predaceous on invertebrates.

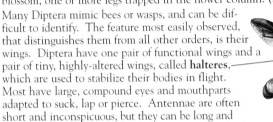

SUBORDER BRACHYCERA

Members of the suborder Brachycera are flies with short antennae—usually with five or fewer segments. Horse and deer flies as well as several predatory and parasitic flies are included in this suborder.

Family Asilidae

Robber Flies

PRED

252. Robber fly of the genus *Laphria*

Robber flies are common, fast-flying predators. The adults prey on flies, bees, grasshoppers, butterflies and beetles, catching them on the wing and landing to suck the hemolymph from their victims. They have keen vision, are very wary, and often attack prey larger than themselves.
Robber flies have thick legs and a relatively large thorax.

253. Unidentified robber fly

Identifiers: Top of the head hollowed out between eyes; body variable: some robust and hairy, resembling bumble bees, others slender and nearly hairless, resembling damselflies; face is usually mustached.
Life Cycle: Larvae occur in rotting wood and soil where they prey on the larvae of other insects.
Range: Over 983 North American species.   **Size:** 5-30 mm (1/4-1 1/8").

Family Dolichopodidae
Long-legged Flies

These are small- to medium-sized predators. They are common in marsh and

254. Long-legged flies

PRED

meadow communities where they can be seen running over leaves in search of their prey, which consist mostly of smaller insects. Males of this family often perform elaborate mating dances.

Identifiers: Usually bright metallic green or coppery bodies; long legs; males often with large genitalia usually folded under abdomen.
Life Cycle: Predaceous larvae live in the soil.
Range: 1227 North American species.
Size: Less than 9 mm (3/8") .

Family Syrphidae
Syrphid or Hover Flies

This is a large group of common flies with extremely variable life histories. Most are wasp or bee mimics. Adults can often be found hovering over flowers. Their bodies can be stocky and covered with hair or brightly striped yellow and black or deep brown.

Identifiers: Large eyes that appear to cover the entire head; proboscis short and fleshy; all are good flyers and often hover.

Life Cycle: Larvae of some species are aquatic, others are scavengers in the soil, but many are important aphid predators.

Range: 874 North American species.

Size: 5-18 mm (1/4-3/4").

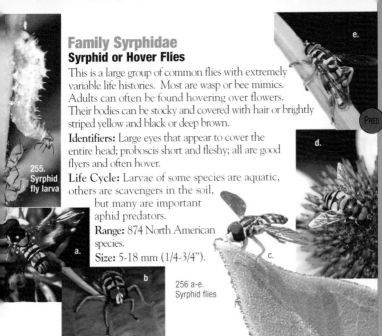

255. Syrphid fly larva

256 a-e. Syrphid flies

PRED

Family Conopidae
Thick-headed Flies

Adult thick-headed flies are found on flowers. They look much like thread-waisted wasps and often have darkened wings. In addition to nectaring, the females wait for appropriate hosts for their larvae. They seize bees and wasps in mid-air, depositing eggs on the surface of their bodies.

Identifiers: Head slightly wider than thorax; abdomen narrowed at base; long, slender proboscis often elbowed.

Life Cycle: Larvae are **parasitoids** of bees and wasps. They ultimately kill their hosts and pupate in the ground.

Range: 67 North American species.   **Size:** 10-13 mm (3/8-1/2").

PRED
NECT

257 a-c. Thick-headed flies

69

HERB
NECT

258. Mating pair of fruit flies, *Rhagoletis* sp.

Family Tephritidae
Fruit Flies

These small flies are often brightly colored or have patterned wings. Many are found on flowers.

Identifiers: Small, with patterned wings; rhythmically wave wings up and down when walking.

Life Cycle: Larvae feed on flowers and fruit or form galls in plant stems.

Range: 280 North American species.

Size: 4-9 mm (1/8-3/8").

• • • • • • • • • •

259. Fruit fly. Notice the spider mimicry enhancing wing pattern.

NECT
PARA

Family Tachinidae
Tachinid Flies

Tachinidae is the second largest family of flies in North America with approximately 1300 species. Most look like houseflies—some are a bit larger. Tachinid larvae are **parasitoids** on a wide variety of insect hosts. Several species of tachinids parasitize caterpillars, including monarchs.

260. Tachinid fly head

261. Tachinid fly

Identifiers: Abdomen generally has a number of very large bristles in addition to the smaller ones; a row of bristles on the thorax above the base of the hind leg and another row just under the base of the wing.

262. Tachinid flies that parasitize monarchs lay their eggs on the outside of a caterpillar. The larvae burrow inside and consume the caterpillar as they grow. They leave the dead host when it is a late instar caterpillar or in the chrysalis stage—lowering themselves to the ground on silk threads like those hanging from the chrysalis to the right.

Life Cycle: Female tachinids lay their eggs on the bodies of other insects. The larvae are **parasatoids**, burrowing into the host and living as internal parasites—eventually killing their hosts. Most drop to the ground and pupate in the soil. Some pupate inside their hosts.

263. The tachinid larva (maggot) and pupae below all developed inside one monarch caterpillar. Normally, they pupate in soil.

Range: 1300 North American species.

Size: 3-14 mm (1/8-1/2").

264. Honey bee approaches *A. speciosa* blossom

ORDER HYMENOPTERA

Many Hymenoptera are active participants in the milkweed community. Their name comes from the Greek word "hymenopteros" meaning "membrane-winged." The name also alludes to Hymeno, the Greek god of marriage, referring to the manner in which Hymenopteran wings are "joined" together in flight by tiny hooks called **hamuli**. Hymenoptera are hard-bodied, active insects, usually with long antennae. Females have well-developed ovipositors. In the case of bees, ants and social wasps the ovipositor has evolved into a stinger. They have chewing mouthparts, but some also have a tongue-like sucking structure. Hymenoptera go through complete metamorphosis.

265. Chewing and sucking structures of a hornet's mouth

The Hymenoptera are divided into two suborders: the primitive, plant eating sawflies, suborder Symphyta; and the wasps, bees and ants of the suborder Apocrita—which are often regarded as the most highly evolved insects. In addition to the highly developed honey bee and ant societies, there are many species of solitary bees and wasps.

Definitive identification of these insects often requires careful observation of wing cells, placement of hairs, apical spurs and other details. The information provided here is fairly general, and more detailed field guides are required to identify these insects to the species level.

SUBORDER SYMPHYTA

Members of the suborder Symphyta— sawflies and horntails—can be seen on flowers or near their host plants. Symphyta larvae are herbivores and resemble lepidopteran caterpillars, but with seven pairs of prolegs and no crochet hooks. They are

(HERB)

266. Sawfly, *Dolerus similis*

known for the damage they do to trees and shrubs. Though adults may nectar on milkweed blooms, no members of this suborder use milkweed as a host plant.

SUBORDER APOCRITA

Members of the suborder Apocrita—the wasps, bees and ants—are some of the most significant members of the milkweed community. There are a great number of species in this suborder. A common feature is their thin waists. The first segment of the abdomen is fused to the thorax and the second and third segments are narrow and form a waist, called a **pedicel**, which gives these insects increased flexibility.

Larvae of the Apocrita are maggot-like, but have a well-developed head with chewing mouthparts. Many are **parasitoids**—they live on or in the bodies of other insects or spiders, eventually killing their hosts. This suborder is divided into 11 superfamilies and can require considerable study to understand. This field guide covers superfamilies that are significant to the milkweed community, common and/or relatively easy to identify. The predaceous stinging wasps of the superfamilies Scolioidea, Vespoidea and Sphecoidea are grouped together and the ants, also of superfamily Scolioidea, are treated separately.

NECT

PARA

267. *Above:* The braconid, *Lysiphlebus testaceipes*, emerging from a parasitized aphid.
268. *Below:* Parasitized aphids, called "aphid mummies," turn brown. *L. testaceipes* pupates inside and emerges from a neat circular hole cut in the top of the dead aphid.

SUPERFAMILY ICHNEUMONOIDEA
Parasitic Wasps

Family Braconidae
Braconids

Braconids are small, usually black or red wasps with thread-like antennae. Adults are nectivores, but the larvae are parasitoids of a great variety of insects and spiders. In the milkweed community, mummified aphids are a fairly common sign that braconids, like *Aphidis colemani* or *Lysiphlebus testaceipes*, are present.

Identifiers: Antennae long but never marked with white or yellow as in the ichneumonids.

Life Cycle: Female braconids lay eggs on a host's body. The larvae burrow into host and feed on its tissues, ultimately killing it. Some species pupate in silken cocoons on the outside of the host's body.

Range: 1937 North American species.

Size: 2-15 mm (1/16-5/8").

• • • • • • • • • •

Family Ichneumonidae
Ichneumons

Ichneumonidae is the largest family of insects. They vary greatly in size and color—many are yellowish to black, while others are brightly patterned. The middle segments of the long, constantly moving antennae of many ichneumons are yellowish or white. Many have long slender abdomens that thicken toward the tips. In addition, most females have long, thread-like ovipositors. Most cannot sting, but a few have short, sharp ovipositors able to pierce the skin. Ichneumons are usually larger and thinner than braconids. Adults drink water and nectar. Larvae are parasitoids on a wide variety of insects and spiders. This family has been divided into a number of subfamilies and tribes which are often parasitic on a particular group of insects.

269 a-c. Rapidly moving antennae, with characteristic yellow or white middle segments, help to identify these insects as ichneumonidae.

Identifiers: Slender, wasp-like; antennae usually at least half as long as body, often with middle segments yellow or white.

Life Cycle: Larvae are parasitoids to specific insects or spiders.

Range: 3322 described North American species. **Size:** 3-75 mm (1/8-3").

- -

SUPERFAMILY CHALCIDOIDEA
Chalcids

The Chalcids are a large group of tiny, difficult to identify wasps, but their impact on any arthropod community is significant. They are blue, blue-black or green. Many are wingless, but those with wings hold them flat over their backs at rest. Most larvae of this group are parasitic or **hyperparasitic** (feeding on other parasites). Many adults are nectivores.

270. Some Chalcids have greatly swollen hind femora.

Identifiers: Short, elbowed antennae; short ovipositor; greatly reduced wing venation; and small size.

Life Cycle: Parasitic or hyperparasitic larvae.

Range: 102 species in North America.

Size: Less than 5 mm (1/4").

© Mike Quinn

- -

NECT

PARA

SUPERFAMILY EVANOIDEA
Family Gasteruptiidae

271. Gasteruptiid

Insects of the superfamily Evanoidea have a unique appearance because their abdomens are attached high above the hind coxae (base segment of an insect's leg). Members of the family Gasteruptiidae are common around flowers.

Identifiers: Similar to the ichneumonidae except for their slender neck; abdomen attached high on the thorax; ovipositors often as long as bodies.

Life Cycle: Gasteruptiid females deposit eggs in the nests of solitary wasps and bees. The larvae eat the eggs and stored food.

Range: 15 species of the family Gaserupidae in North America.

Size: 13-20 mm (1/2-3/4").

SUPERFAMILY PELECINOIDEA

There is only one species of Pelecinoidea in North America, *Pelecinus polturator*, the American pelecinid. Females are distinctive with long antennae and an

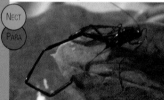

NECT

PARA

extremely long slender abdomen. They are common nectivores often found in crop fields and suburban gardens.

Life Cycle: The female shoves her abdomen deep into the soil to lay eggs inside the bodies of May beetle larvae and, perhaps, other scarabs.

Range: 1 species in eastern North America.

272. American pelecinid, *Pelecinus polyurator*   **Size:** 45-50 mm (1 3/4-2").

SUPERFAMILY BETHYLOIDEA
Cuckoo Wasps Family Chrysididae

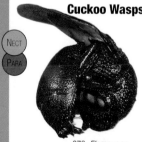

NECT

PARA

Cuckoo wasps are common nectivores with brilliant blue and green metallic coloring. They do not sting.

Identifiers: Body metallic blue or green with coarse sculpturing; curl into a ball when disturbed.

Life Cycle: Larvae are parasites and live in the nests of other wasps or bees.

Range: 227 North American species.

273. *Elampus* sp.   **Size:** 6-12 mm (1/4-3/4").

74

STINGING WASPS

The wasps of the superfamilies Scolioidea, Vespoidea and Sphecoidea are stinging wasps. They have long (12-13 segmented) antennae. Females sting and their larvae usually feed on animal food provided by the adults. Many of these wasps nectar on milkweed flowers.

PARASITIC STINGING WASPS—SUPERFAMILY SCOLIOIDEA

Scolioid wasps are parasitoids. Females are often wingless and ant-like.

Velvet ants, Family Mutillidae, are 6-25 mm (1/4-1") long and brightly patterned with red, yellow and orange. Males have wings, but females are wingless and look like ants except for their hairy antennae. They also lack the bead-like connecting abdominal segment present in ants. Most velvet ant eggs are laid in the nests of ground-nesting wasps and bees where they kill the host's larvae and live off the food brought by the adult host. A few parasitize beetles and flies. Females can inflict a painful sting which has earned them the name cow-killers.

274. Preserved specimens of the velvet ant species *Dasymutilla asopus*—male above, female below.

Range: 483 species in North America.

275. Scoliid wasp, *Campsomeris tolteca*

Scoliid wasps, Family Scoliidae are 10-35 mm (1/8-1 3/8") long, robust, hairy, dark wasps with light markings. A female scoliid wasp searches in the ground for beetle larvae. Once she finds one, she stings it, builds a small chamber in the ground around the paralyzed host and lays an egg on it.

Range: 483 species in North America.

Tiphiid wasps, Family Tiphiidae, are black or yellow and black, medium-sized wasps. Their abdominal segments are separated by strong constrictions. The tiphiid wasp, *Myzinum quinquecinctum*, lays its eggs on May beetle larvae. The wasp larvae eat the host's non-essential tissues first, but eventually kill it.

276. *Myzinum quinquecinctum*

Range: 225 species in North America.

PREDACEOUS STINGING WASPS

Wasps from the superfamilies Vespoidea and Sphecoidea are predaceous wasps. They generally build nests for their eggs, capture prey, process it and stock their nests with food for their larvae.

SUPERFAMILY VESPOIDEA
Vespoid Wasps

277. Paper wasp, *Polistes* sp.

Vespoid wasps include the long-legged, dark-colored spider wasps of the family Pompilidae and the dull brown and black or yellow and white banded wasps of the family Vespidae.

Spider Wasps, Family Pompilidae

PRED
NECT

Spider wasps are 10-50 mm (3/8-2") long. They hunt only spiders. Adults often walk on the ground flicking their wings. Females build a special cell or burrow in the ground. They then capture and paralyze spiders on which they lay their eggs.

Range: 288 species in North America.

278. Spider wasp, *Sericopompilus apicalis*

Vespid Wasps, Family Vespidae

PRED
NECT

Vespid wasps feed their developing larvae chewed insects of various sorts. They have distinctly notched eyes and a very long **discal cell** on their front wings. The family includes seven North American subfamilies, including the familiar potter wasps, yellow jackets and hornets, and paper wasps.

Range: 415 species of Vespidae in North America.

Potter wasps, Subfamily Eumeninae, are 10-20 mm (3/8-1") long, black wasps with yellow or white markings. These wasps are solitary. The female builds a round chamber of mud on a twig into which she seals her larvae and an anesthetized caterpillar or sawfly larva as food.

Yellow jackets and hornets, Subfamily Vespinae, are social insects. Adults eat nectar, juice from fruits and perhaps other insects. They feed their larvae pre-chewed insects. Nests are begun in the spring by a queen and added onto as workers are produced. Hornets build tiers of cells of a papery material surrounded by an outer covering. Yellow jackets build their nests below ground. Females inflict painful stings.

279a. Hornet

279b. Yellow jacket, *Vespula maculifrons*

Paper wasp females, Subfamily Polistinae, work cooperatively to construct a singular, roughly circular, tier of cells attached to

280. Paper wasp attacks a monarch caterpillar

PRED
NECT

the underside of a surface such as a porch ceiling or eaves. They drink nectar or juices from crushed or rotting fruit and feed their larvae pre-chewed insects. Paper wasps collect caterpillars and chew them up into small portions to feed their larvae. These wasps may have a significant impact on populations of caterpillars, including monarchs. There is some evidence that, though they will regularly feed on monarchs, they prefer other caterpillars.[6] Medium-sized monarch caterpillars, from late third instar through early fifth instar, are most at risk for paper wasp predation.

SUPERFAMILY SPHECOIDEA
Family Sphecidae, Sphecid Wasps

Sphecid wasps are a large and diverse family of solitary hunters ranging from 10-55 mm (3/8-2 1/8") long. They may be solid black or brown or patterned with red, yellow and white. They nest in the ground, in natural cavities or make mud nests. Adults drink nectar, aphid honeydew or the body fluids of prey. The adults are often found on flowers. Each species provisions its nest with characteristic prey species.

PRED
NECT

Range: 1139 North American species— divided into 9 subfamilies.

Mud dauber females, *Sceliphron* spp., build adjacent tubular cells. They stuff one paralyzed spider into each cell. After laying one egg on the spider, they seal the cell. Adults feed on nectar.

281. Mud dauber, *Sceliphron caementarium* (preserved specimen)

[6] Rayor, L.S. 2004. Effects of Monarch Caterpillar Host-plant and Size on *Polistes* Wasp Predation. In: **Monarch Butterfly Ecology and Behavior**, K. Oberhauser & M. Solensky, Eds, Cornell University Press. (In press)

282. Thread-waisted wasp

Thread-waisted wasps, *Ammophila* spp., feed on nectar. Females dig a short burrow with an enlarged chamber at the end. They drag an immobilized caterpillar or sawfly larva on which they have laid an egg into the chamber. They then seal the burrow temporarily, but may return to bring more food to the larva. A female will remember and tend several nests.

Golden digger wasps, *Sphex ichneumoneus*, nectar on milkweed and may also drink fluids from the crickets and long-horned grasshoppers they collect to feed their larvae. The female digger wasp digs a nearly vertical tunnel in hard packed soil. The tunnel ends in two to seven individual underground cells. She places one anesthetized prey, with one egg laid on it, in each cell.

283. Golden digger wasp, *Sphex ichneumoneus*

Superfamily Scolioidea
Family Formicidae
Ants

Ants are familiar, 1-25 mm (1/16-1") long, and mostly black, brown or reddish insects. They can be distinguished from mimics and other insects by their strongly elbowed antennae and the bead-like formation at the waist between their thorax and abdomen. Each species has a slightly different, complex social structure. Usually workers are sterile, wingless females. Ants live in colonies underground or in dead wood. Periodically reproductive, winged males and females emerge from the nest to mate. After the mating flight, males die and females lose their wings and return underground to start new colonies.

Most ants in North America come from two subfamilies: the Myrmicinae and the Formicinae.

The Myrmicinae sting. Their first two abdominal segments are bead-like. This subfamily includes fire ants (*Solenopsis* spp.) and harvesters.

The Formicinae do not sting and have only one bead-like abdominal segment. They include carpenter ants, mound building ants and field ants.

PRED
HERB
DEC/S
NECT

284. Fire ant, subfamily Myrmicinae

285. Subfamily Formicinae

Ants are major players in the milkweed community. They can be predators, nectivores, scavengers, and occasionally herbivores and farmers as well.

Fire ants, found in Florida and the Gulf states, west to the

286. Ants scavenging the carcass of an insect that fell victim to an ambush bug on the milkweed flower above.

Pacific coast and north to British Columbia, are notorious predators, killing many monarch caterpillars. Mound builders scavenge the honeydew left on the surfaces of the milkweed leaves by aphids. Many kinds of ants sip nectar from milkweed blooms. Although no ants are known to tend the yellow oleander aphids common on milkweed, they often tend black and green aphids on these plants. Possibly to protect "their aphids" from potential predators and competitors, they remove and kill many other insects, including monarch eggs and larvae, from plants.

SUPERFAMILY APOIDEA
Bees

Bees comprise a large group of insects adapted to feed at flowers, gathering nectar and pollen. More than 3500 species occur in North America. They can be 4-25 mm (1/8-1") long, black or brown or banded with yellow, white or orange, and covered with branched, feathery hair. They are important pollinators in any natural community and can be found in great

287. Honey bee

variety and abundance on flowering milkweed. The branched hairs and pollen comb on their legs are used for collecting pollen, but are less useful at milkweed blooms. (*See page 16.*) However, bees are successful milkweed pollinators and carry many milkweed pollinia between plants.

Most bees are solitary, with single females nesting in the ground or in natural cavities and caring for their own offspring. Bumble bees and honey bees are social, living in colonies that consist of a fertile queen, sterile female workers, and males called **drones**. The social bees are the only ones that produce honey. Cuckoo bees lay their eggs in the nests of other bees.

Bees are classified by differences in their wings and tongue length. This can be challenging for beginning observers. The bee tongue, when not in use, is folded up tightly under the head and comparison of wing venation can be difficult in the field. Other clues, though not as specific for absolute species certainty, can prove useful for understanding the interactions of bees in the milkweed community.

SHORT-TONGUED BEES
Family Colletidae
Yellow-faced and Plasterer Bees

NECT

Yellow-faced bees, *Hylaeus* spp., are common, 5-6 mm (1/4") long, slender, black bees. The abdomen is wasp-like and nearly hairless. There are yellow markings on their faces, pronotums and tibiae. Most species make silk lined brood cells in the pith of stems and stock them with nectar and pollen.

Plasterer bees, *Colletes* spp., are less common than yellow faced bees, densely covered with hairs and brown with whitish to yellowish bands

288. Plasterer bee, *Colletes* sp.

on their abdomens. Their burrows look like low mounds of dirt. At the end of an entryway a few inches deep, the female digs branching tunnels with individual brood cells. She lines her burrows with a thin layer of saliva that dries to a cellophane-like consistency. Once a cell is stocked with pollen, nectar and one egg, it is sealed shut to protect the larvae from predators and parasites.

Family Halictidae
Halictid or Sweat Bees

Small, 5-15 mm (1/4-5/8"), halictid bees often nest close together underground, sharing the same passageway. Some have bright, metallic coloring.

Range: 502 species in North America.

289. Halictid bee

Family Megachilidae
Leafcutting Bees

Leafcutting bees are stout, dark-colored bees, 10-20 mm (3/8-1") long. Several species are common. They have special hairs on the underside of their abdomens called "**scopa**" that are used to carry pollen. They nest underground or in natural cavities and make cell partitions with mud, resin or leaf pulp.

Range: 682 species in North America.

290. Leafcutting bee, family Megachilidae

LONG-TONGUED BEES
Family Apidae
Digger Bees, Cuckoo Bees, Carpenter Bees, Bumble Bees, and Honey Bees

This is a large group of bees that varies greatly in size, appearance and habits. There are 3 subfamilies: Anthophorinae, Xylocopinae, and Apinae. Members of the subfamily Apinae are the social bumble bees and honey bees. The other two subfamilies are solitary. Each subfamily is further divided into tribes. Only a few common representatives are covered in this text.

Range: 977 species in North America.

The Subfamily **Anthophorinae** includes the **digger bees** and the **cuckoo bees**.

Digger bees are robust and hairy bees, 10-20 mm (3/8-1") long. They often nest together and are called "flower-loving bees" because of the great variety of flowers that they visit.

291. Digger bees, *Melissodes agilis*. Female above, male below. (preserved specimens)

Cuckoo bees are relatively bare and wasp-like in appearance. They enter the brood cells of other bees and lay their eggs next to the hosts' larval food supply. The cuckoo bee larvae hatch first and eat all of the host larvae's food.

292. Cuckoo bee, *Nomada maculata* (preserved specimen)

Carpenter bees, Subfamily Xylocopinae are large, robust and resemble bumble bees. They can be distinguished from the bumble bee by their shiny, black, hairless abdomen. These bees make their nests in wood or in the pith of stems of various bushes.

The Subfamily **Apinae** includes the familiar bumble bees and honey bees.

Bumble bees, Tribe Bombini, *Bombus* spp. are common, robust, black and yellow, hairy bees, usually 15-25 mm (5/8-1") long. Most are social. Their colonies contain two generations—the queen and her offspring. The queen lays eggs, feeds larvae and maintains the hive, while female workers bring in loads of pollen and nectar. Colonies are annual; new queens are produced at the end of every year. These

293. (*above*) and 294. (below) Carpenter bees, subfamily Xylocopinae

queens overwinter and start new colonies in the spring.

Honey bees, Tribe Apini, *Apis mellifera*, are well-known and common insects. Only one species of honey bee occurs in North America, although this species includes several genetic lines that vary slightly in color. Honey bees were first brought to North America in the 1600's by European settlers. The original strains have recently been joined by the Africanized honey bee. The two subspecies are nearly identical in appearance and their stings are similar. However, Africanized honey bees can be quite different in their behavior. They are often very protective of their hives. Sometimes, when the hive is disturbed, they will exit en masse and give chase 300 meters or more, earning them the popular name "Killer Bees." The subspecies can interbreed and all southern feral (wild) honey bees are likely Africanized to some degree.

Most European honey bees live in man-made hives or hollow trees and other cavities. They produce honey and beeswax and are excellent pollinators. Honey bees have a complex social system that revolves around maintaining the queen bee for 2 to 5 years. The queen produces a colony of 60,000-80,000 workers that collect nectar and pollen, store honey and maintain the hive. Workers feed the queen and larvae and control the development of new queens. New queens, fed as larvae on the same royal jelly offered to the hive queen, are produced in late spring or early summer. The old queen departs with a swarm of workers to establish a new colony. A hive can produce more than one new queen and release more than one swarm. Within seven to ten days of emerging, new queens leave the hive to mate. Males, called drones, die after mating. At the end of the swarming episode, only one queen will remain in the original hive.

295. (*above*) and 296. (*left*) Bumble bees, *Bombus* sp(p).

297. Honey bee, *Apis mellifera*

298. Honey bee, *Apis mellifera*. Note the worn, tattered wings.

83

CLASS ARACHNIDA
Arachnids

299. Orb weaver spider, family Araneidae, feeding on a queen butterfly

Arachnids are wingless arthropods with four pairs of legs. Spiders, Order Araneae, are the most familiar arachnids, but harvestmen, Order Opiliones, and mites, Order Acari, are also regular members of the milkweed community.

ORDER ARANEAE
Spiders

Spiders are extremely common predators with 3000 named species in North America and many more as yet unnamed. A great variety of spiders can be found in a milkweed patch.

Most spiders produce venom to immobilize or kill their prey, but in North America only two species are dangerous to humans, the brown recluse and the black widow.

Spiders inject their prey with digestive juices containing enzymes that dissolve the prey's body tissues. The spiders then suck the resulting liquid.

Spider eggs are laid inside a silken sac. Most of the sacs are thick and fluffy to protect the spiderlings. Many spiders carry their egg sacs until the young spiderlings hatch, and some species care for their young until the first molt.

300. Green lynx spider, *Peucetia viridans*, tending young

84

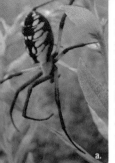

301a. Orb weaver spiders from the family Araneidae and the genus *Argiope*, like the garden spider, *Argiope aurantia*, to the left, are common in milkweed patches. b. These spiders incorporate a zigzag pattern in their webs like the ones so clearly visible in the web of the tropical *Argiope* to the right. c. Look for the zig zag in the web below (*left*).

Family Araneidae
Orb Weavers

There are several hundred species of Araneidae spiders in North America. These spiders are 2-28 mm (1/16-1 1/8") long. They often have large abdomens, occasionally with strange angular shapes. Their eyes are arranged in two horizontal rows of four eyes each. They have poor vision and depend on the vibrations from their webs to alert them to the presence of prey. The symmetrical web of this family with its central hub and radiating spirals is so recognizable it has become our symbol for all spiders. The webs may be spun vertically, horizontally, or on an angle. Nocturnal orb weavers build their webs at dusk. Diurnal orb weavers, like those of the genus *Argiope*, build their webs at dawn. Zigzag patterns or "stabilimenta" incorporated in the webs of *Argiope* spiders appear to act as warnings to keep birds from flying through the webs. Spiders from a few other families also spin orb webs.

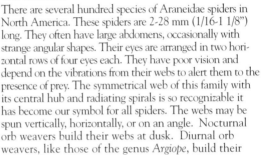

302. This orb weaving spider from the family Tetragnathidae, the long jawed spiders, was found on milkweed near woodlands.

Family Pisauridae
Nursery Web Spiders

Nursery web spiders are often found in milkweed stands. They are large, 7-26 mm (1/4-1") long, and long-legged. Their eyes are arranged in two, u-shaped rows. They are hunters and do not spin webs to catch their prey. The female carries her egg sac in her jaws until it is nearly ready to hatch. She then suspends it between leaves tied together with silk and stands guard nearby. The leaves of the broad-leaved varieties of milkweed, such as *A. syriaca*, are well-suited to this task. It is not unusual to find female nursery web spiders guarding their egg sacs in the milkweed patch.

303. Nursery web spider standing guard on a milkweed stem below her egg sac

85

Family Thomisidae
Crab Spider

Crab spiders stand with their legs outstretched and ambush passing insects. Their stance and movements are much like those of

304. *Misumena* sp.

crabs. They are frequently found nestled in among milkweed flowers. They can detect even very slight movement and often catch insects much larger than themselves. Crab spiders can change color to match their surroundings. The same species may look pink on a common milkweed blossom and yellow on a goldenrod blossom.

305. Crab spider

• • • • • • • • • •

Family Salticidae
Jumping Spiders

Jumping spiders are small (less than 15 mm (3/4") long), compact spiders, often with brightly colored patterns. Two of their eight eyes are quite large. They have excellent sight and can recognize prey from 10-20 mm (4-8") away. Jumping spiders leap onto their prey from many times their own body length away and capture prey much bigger than themselves, including monarch caterpillars. They secure themselves with a silk thread before jumping in order to be able to climb back if they miss.

306. Daring jumping spider, *Phidippus audax*

ORDER OPILIONES
Harvestmen
Family Phalangiidae
Daddy-long-legs

There are between 100 and 150 species of Phalangiidae in North America. Daddy-long-legs are often mistaken for spiders. They have four pairs of long, thin legs and a broad, oval body consisting of a compact cephalothorax joined to the

307. Daddy-long-legs drinking nectar

abdomen. Unlike spiders, daddy-long-legs produce neither silk nor venom. They hold their bodies fairly close to the ground—their long stilt-like legs bent well above their bodies. Their two eyes are raised on short stalks in the middle of the cephalothorax and point sideways. They can distinguish light and dark, but not images. They receive most of their sensory information from their long legs, especially the second pair.

Most daddy-long-legs are active at night, but they can be seen cruising the milkweed patch at most any time. These creatures are opportunists. They eat dead and decaying plant and animal matter; hunt and capture small animals such as flies, spiders, aphids, snails and earthworms; and drink plant juices and nectar.

308. Daddy-long-legs on *A. syriaca*

ORDER ACARI
Ticks and Mites

Family Trombididae
Velvet Mite

Trombidium spp.

Mites are the most abundant and diverse arachnids on the planet with over 30,000 species, but because of their small size they are often overlooked. Velvet mites are predators and parasites in the milkweed community. Adults are 3-4

mm (1/8") long, covered with dense, velvety red hair and have eight legs. They feed on insect eggs, including those of the monarch butterfly. Mite eggs are scattered in the soil. They go through a complex life cycle being pupa-like in the first stage, active and parasitic on insects and other arthropods in the second stage, and still and pupa-like again in the third stage before becoming adults. Immature mites can be mistaken for insects because they have only six legs, but they lack the three distinct body parts of insects. Parasitic mite nymphs can be found on aphids.

309. A red velvet mite eats a monarch egg.

87

GLOSSARY

abdomen - the elongate hind part of the body, behind the thorax.

abiotic - the non-living components of an ecosystem.

adult - the final stage in metamorphosis, after the pupa.

antenna (plural, antennae) - sensory organ on the head of an insect.

anther - the organ at the upper end of the stamen that secretes and discharges pollen.

apical - on the highest point of something.

ballooning - a behavior in which larvae or spiders produce long threads of silk which can carry them long distances on the wind.

biomass - the total mass of all living things or the mass of a particular species.

cephalothorax - the fused head and thorax region found in some arthropods.

cerci - paired appendages on the end of the abdomen of many insects that are used for sensing, defense or mating.

chelicera (plural, chelicerae) - one of the first pair of the six pairs of appendages found on the anterior part of an arachnid. These may be pinching or piercing in structure and used in defense or digging.

chloroplasts - organelles that contain chlorophyll and are used in photosynthesis.

chrysalis (plural, chrysalides) - another name for a butterfly pupa.

class - a taxonomic category following kingdom and phylum and preceding order.

climax (community) - the final stage of plant succession in which equilibrium is achieved.

clone - a group of genetically identical cells or individuals produced from a common ancestor.

communal feeders - a group of animals, often related to each other, that feed together.

competitors - individuals of the same species or different species that vie for the same limited resources.

compound eye - the large, multifaceted eye of an insect.

cornicle - a tubular structure on each side of the abdominal regions of aphids from which pheromones or honeydew are excreted.

corpusculum - a clip-like structure on top of the pollinarium to which the pollen-filled pollinia are attached.

coxa - the basal segment of the leg.

cuticle - a layer of the body wall in invertebrates which protects against desiccation, mechanical or chemical damage.

decomposer - an organism which feeds on decaying matter, breaking it down into nutrients.

diapause - a period of dormancy between periods of activity in insects, usually induced by environmental signals.

discal cell - large cell (space between veins) in the center of a wing near its base.

division - a taxonomic category in the plant kingdom that corresponds to phylum in the animal kingdom.

drone - the male of ants, bees and wasps, whose function is to mate with fertile females.

eclose - to emerge from the pupal stage.

elytron (plural elytra) - the thick, leathery forewing of beetles, earwigs and some Homoptera.

exoskeleton - the skeleton on the outside of an invertebrate's body that protects it and serves as a point for muscle attachment.

family - the taxonomic category between order and genus.

femur (plural, femora) - the largest segment of a leg.

flower column - the structure which holds the reproductive parts of the flower.

flower stalk - the stem that holds an individual flower.

follicle - a small sac, cavity or gland. The botanical term for a milkweed pod.

food chain - the transfer of energy from producers through a series of organisms that eat and are eaten.

food web - a diagram that shows the feeding relationships of organisms in an ecosystem.

forb - any broadleaf, herbaceous plant.

genus (plural genera) - a group of closely related species; the first name in a scientific name, capitalized and italicized when written.

grub - a thick, worm-like larva of some beetles and other insects.

habitat - the place in which an organism or community lives, characterized by its biotic and abiotic properties.

hair pencils - hairs which extend from the abdomen of male Lepidoptera and may transfer pheromones during courtship.

haltere - a highly modified hind wing in Diptera that is used as a balancing organ, enabling the fly to sense movement and direction in flight.

hamuli - a row of hooks along the edge of the hind wing which attach to a fold in the forewing of Hymenoptera.

head - a structure which is anterior to the thorax in arthropods.

hemolymph - the name for the blood of insects.

herbivore - an animal that feeds on plants.

honeydew - plant sap that has passed through the bodies of aphids.

hood - a modified anther on milkweed enclosing a horn containing nectar.

host - the plant or animal on which an herbivore or parasite lives and feeds.

host specific - referring to herbivores or parasites, adapted to feed only on a specific group of organisms.

hypermetamorphosis - insect development where successive instars have quite different forms.

hyperparasitic - a parasitic insect that lives in or on a host that is parasitic to another insect.

instar - a period between larval molts.

kingdom - a category in classification that contains related phyla.

larva (plural, larvae) - the stage in the life cycle of an insect after the egg in metamorphosis. Lepidopteran larvae are called caterpillars.

maggots - fly larvae lacking a distinctive head, usually found in decaying matter or as a parasite.

mandibles - strong "jaws" on an insect head.

mesothorax - the second thoracic segment of an insect which carries a pair of legs and possibly wings.

metamorphosis - a change in form during development; the transformation of an organism from the larva to the adult stage.

  complete metamorphosis - transforming through the stages of egg, larva, pupa and adult.

  simple metamorphosis - transforming through the stages of egg, nymph and adult.

metathorax - the third thoracic segment of the insect which carries a pair of legs and possibly wings.

molt - the process of shedding the skin or exoskeleton.

mouthparts - organs involved with feeding. In insects they are made up of labrum in front, a hypopharynx behind the mouth, a pair of mandibles and maxillae and a labium forming the lower lip.

natural succession - a progressive series of changes in a community from the initial colonization to the climax community.

natural community - the intersection of many habitats.

naturalize - allowing a plant community to develop without disturbance.

nectivore - animal that feeds on plant nectar.

niche - the function of an organism in its environment, comprising the habitat, time of activity and resources it obtains there.

nymph - an immature stage of incomplete metamorphosis between egg and adult.

obligate parthenogenetic - a species which always reproduces asexually.

ocelli - simple eyes of some insects.

order - the taxonomic category between class and family.

organelle - an organ found inside of a cell such as a nucleus or chloroplast.

ovum (plural, ova) - unfertilized egg cell.

ovipositor - specialized egg-laying organ on the abdomen of most female insects.

parasite - organism that lives in or on a host's body and depends on the host for nutrients and resources necessary to complete its life cycle. Parasites do not usually kill the host but may weaken it and make it more susceptible to disease or predation.

parasitoid - an insect that lays eggs on or inside another insect species (the host). The eggs hatch and larvae feed on the host from the inside, eventually killing it.

parthenogenesis - development of an individual from an egg without male fertilization.

passersby - organisms that use the plant as a resting place but do not feed off the plant material directly or eat invertebrates typically found on the plant.

pedicel - the short, narrow portion of the body that connects the abdomen with the cephalothorax of a spider; the short narrow part of the abdomen that connects to the thorax in ants; the second segment of the antenna; a flower stalk.

pedipalps - the second of the six pairs of appendages in arachnids. They are tools for the killing and manipulation of prey and used for defense and digging. They also serve as tactile and olfactory organs.

phylum (plural, phyla) - a taxonomic category after kingdom and containing related classes.

pollinarium - the flower structure

that contains the plant's pollen. It is comprised of the corpusculum, rotator arms and pollinia.

pollinia - the waxy, pollen-filled structures of a flower.

predator - an organism that obtains energy by consuming, usually killing, another organism (prey).

proboscis - an extended structure of the mouthparts through which some insects draw liquid.

pronotum - a plate-like area on the front and top of the prothorax.

producer - an organism that synthesizes its own food, a green plant.

prothorax - the first thoracic segment of an insect that bears a set of legs.

pupa (plural, pupae) - the third stage in metamorphosis, after the larval stage.

ramets - plant stems that originate from the same underground rhizome.

raptorial - adapted for seizing prey.

recumbent – lying down.

reproductive diapause - a period of sexual dormancy between periods of activity.

rhizome - a horizontal, underground stem.

rostrum - a piercing-sucking snout or beak.

rotator arm - two thread-like structures that connect the corpusculum to the pollinia in a flower pollinarium.

scavenger - an animal that feeds mainly on other dead animals or animal products.

scientific name - the genus name followed by the species name for an organism, italicized when written.

scopa - a specialized tract of hairs that bees use to compact pollen to take back to the nest.

scutellum - a triangular shaped section on the back of Hemiptera, Homoptera and some Coleoptera.

self-incompatible - plants and animals that only breed with others who are not closely related.

sequester - to separate or segregate something from the general whole without changing it. When monarchs sequester cardenolides, they do not metabolize or break down the chemicals but rather shunt them into specific locations in their bodies.

species - a group of distinct organisms that can interbreed with one another.

spinneret - the organ from which silk is spun.

spiracles - openings on the thorax and abdomen of insects through which gases are exchanged with the outside air. These lead to long air tubes, or tracheae, that run throughout the body.

stigmatic chamber - the part of the flower where pollen is delivered to the ovaries for fertilization.

strategy - a plan of action which aids in an organism's survival.

systematics - the study of relationships among organisms.

tarsus (plural, tarsi) - insect feet made up of several segments that usually end in a pair of claws.

taxonomy - the science of classification.

thorax - the middle section of an insect's body. The wings (if present) and legs are attached to this section.

tibia - the section of a jointed insect leg that is below the femur and above the tarsus.

trochanter - the second joint of the leg of an insect located below the coxa and above the femur.

umbel - a cluster of flowers with individual flower stalks originating from a single axis point.

REFERENCES

There are thousands of excellent references on milkweed, monarchs and the organisms with which they interact. The following were important sources for this book.

BOOKS

Alcock, John. 1997. *In a Desert Garden.* W.W. Norton & Company. New York.

Borror, Donald J. and Richard E. White. 1998. *A Field Guide to Insects: America North of Mexico.* Houghton Mifflin Co. Boston and NewYork.

Hoth, J, I. Pisanty, K. Oberhauser, L. Merino and S. Price, editors. 1999. *Proceedings of the North American conference on the Monarch Butterfly.* Commission for Environmental Cooperation. Montreal, QC.

Malcolm, S. B. and M. P. Zalucki, editors. *Biology and Conservation of the Monarch Butterfly*. Natural History Museum of Los Angeles County, Los Angeles.

McGavin, George C. 1999. *Bugs of the World.* Blandford. UK.

Milne, Loris and Margery. 1980. *National Audubon Society Field Guide to North American Insects and Spiders.* Alfred A. Knopf. New York.

Newcomb. Lawrence. 1977. *Newcomb's Wildflower Guide.* Little, Brown and Company. Boston, Toronto and London.

Scott, James A. 1986. *The Butterflies of North America: A Natural History and Field Guide.* Stanford University Press. Stanford, CA.

Stokes, Donald. 1983. *A Guide to Observing Insect Lives.* Little, Brown and Company. Boston, Toronto and London.

Swan, Lester A. and Charles S. Papp. 1972. *The Common Insects of North America*. Harper & Row. New York, Evanston, San Francisco, London.

Wright, Amy Barlett. 1993. *Peterson First Guide to Caterpillars*. Houghton Mifflin. Boston and New York.

SCIENTIFIC PAPERS AND MAGAZINE ARTICLES

Dailey, Patrick J., Robert C. Graves, and John M. Kingsolver. 1978. Survey of the Coleoptera collected on the common milkweed *Asclepias syriaca*, at one site in Ohio. *The Coleopterists' Bulletin*, 32(3), pp 223-230.

Dailey, Patrick J., Robert C. Graves, and John M. Kingsolver. 1978. Survey of Hemiptera collected on the common milkweed *Asclepias syriaca*, at one site in Ohio. *Entomological News*, Vol. 89, Nos. 7 and 8, pp 157-162.

Woodson, Robert E. 1954. The North American Species of Asclepias L. *Annals of the Missouri Botanical Garden*, Vol. 41 No. 1.

WEBSITES

MonarchLab, University of Minnesota. *http://www.monarchlab.umn.edu.*

Monarch Watch, University of Kansas. *http://www.monarchwatch.org.*

Monarch Larval Monitoring Project, University of Minnesota. *http://www.mlmp.org.*

This index includes the names of arthropods, plants, insect-borne diseases and related subjects covered in this field guide. References to various topics and technical terms may also be found in the Table of Contents and Glossary.